The Uncanny Muse

Also by David Hajdu

The Uncanny Muse

MUSIC, ART, AND
MACHINES FROM
AUTOMATA TO AI

David Hajdu

W. W. NORTON & COMPANY
Independent Publishers Since 1923

Manufacturing by Lake Book Manufacturing
Book design by Daniel Lagin
Production manager: Lauren Abbate

ISBN 978-0-393-54083-3

W. W. Norton & Company, Inc., 500 Fifth Avenue, New York, NY 10110
www.wwnorton.com

W. W. Norton & Company Ltd., 15 Carlisle Street, London W1D 3BS

1 2 3 4 5 6 7 8 9 0

For my teachers Helen Epstein and Michele Wallace,

and in memory of Tad Mosel.

And for my students, who keep teaching me.

Contents

Preface

FACELESSNESS
AND TIME

I got to thinking: Why is this so disturbing? I had seen art of many kinds that had given form to much darker feelings: art of rage and violence. This wasn't like that. This was not quite like any art I had seen before, and I couldn't process how it could involve feelings of any sort.

It was February 2019, and I was at the opening of an art show at HG Contemporary, a trendy space in the Chelsea gallery district of Manhattan. The name of the exhibition was "Faceless Portraits Transcending Time." The artist was an algorithm programmed by Ahmed Elgammal, a professor of computer science at Rutgers University. He had anthropomorphized the program somewhat, calling it AICAN—like "AI can." I had come out of curiosity to see what it could do.

According to the catalog, this was the first show of algorithmic art in a New York gallery, described as "a collaboration between artificial intelligence and its creator." At least thirty people were present, along with Elgammal, a boyish-looking man with a shaved head wearing rectangular horn-rimmed glasses. He stayed mainly in one spot near the entrance to the gallery, standing with his arms crossed, smiling a closed-mouth smile.

I wandered about, gazing at the art and soaking in the atmosphere, which struck me as odd. People moved around the space quickly, instead of lingering in front of the canvases. I saw several people shake their heads and move on. I heard more than one person laugh, though I couldn't tell the reason. As people walked by me, I picked up bits of praise for the art—to be expected at an opening—but it was mixed with grumbling and wise-cracks. I scrawled down notes from the snippets of conversations I over-heard: *Fascinating . . . I'd like to know how this was really done . . . God, it's inhuman . . . Impersonal . . . What's the point? Mechanical . . . Cold . . . This is so weird . . . Are we supposed to believe this is real art?*

I took in the work—abstracted representational portraits done in scratchy gestures and blurry colors, with faces missing—strange, unset-tling work—and tried to process what I was hearing and understand what I was feeling. In the background, "Warszawa," a moody synthesizer track from David Bowie's album *Low*, played softly. I started to feel a bit light-headed.

I got to wondering: How *was* this artwork done? What does it mean for a human being to collaborate with a mathematical formula? Yes, this work *is* fascinating, and also, for me, profoundly disorienting. I couldn't figure out why. Was it the facelessness, the absence of humanity at its heart? I started to ponder if that might be something I could get used to, the way we adjust to new artforms over time. Then again, I wondered, how new is this, really?

What's so bad about being mechanical, anyway? Isn't the music in the background mechanical, too? People love *that*.

Besides, what's so great about being human? Humans are awfully flawed creatures. Aren't the flaws in humanity, our complications, essen-tial to our character, if not attributes? Then, why criticize computer art for being flawed, too? My head was spinning.

At the time of this show, the field of art and music made by artifi-cial intelligence was beginning to expand dramatically, shaking up the way many people thought about creativity and culture. I knew about the strange way of feeling people say they experience in the presence of human-made creations that seem improbably, if not impossibly human: *entering the Uncanny Valley*. I thought, what is this place, and how did we get here?

The Uncanny Muse

MASKELYNE AND COOKE'S AUTOMATA AT THE EGYPTIAN HALL.

1

IS IT ALIVE?

Her name was Zoe. Not *its*: her. The gender was clearly established in etchings on posters for her performances in shows by the magician John Nevil Maskelyne in Piccadilly. Zoe looked the very image of a proper Victorian girl around the age of eleven or twelve. Yet, something elemental—not who she was, but what she was—was not so easy to grasp. When her name appeared in print, as it did frequently in advertisements and newspaper articles, it would often be set within quotation marks, a typographical wink to hint that there was a bit of secret knowledge, something unspoken or unknown or perhaps even unknowable, at work.

A set of studio photographs was taken of Zoe for publicity in the 1880s, when she was one of the most celebrated figures of the day, promoted as "the greatest attraction in London." The photos depict a pretty, dark-haired girl dressed in a luxuriously ruffled dress with fancy lacing on the hem and matching filigree in the cuffs of her flowing, billowy sleeves and high Victorian neckline. We see her in profile, her hair pulled behind her head and twisted into a tight braid, showing off her soft features and satin complexion. She is sitting straight-backed atop a high wooden pedestal to which an artist's easel has been attached for her use. Her gaze is fixed on a second figure in the photo, an adult man with a long, inward-curling mustache, dressed in a black tuxedo and standing rigidly at a three-quarter angle. She stares at him as she sketches his portrait, and he stares into the camera lens as he points to her with his right hand, signaling for us to watch what this girl is doing. Look at her: She's *making art.*

In Zoe's public appearances, the grandly mustachioed man, Maskelyne, would carry her onto the stage in his arms, introducing her to the audience in a vision of paternal caregiving that sweetly evoked Bob Cratchit holding Tiny Tim while obscuring the fact that Zoe's legs were made of wire work and could never have held the weight of the complicated machinery built into her torso, hidden beneath the poofy billows of her dress. Stepping down off the stage, Maskelyne would walk along the aisles of the theater, allowing members of the audience to inspect Zoe for themselves, to touch her and see that, while she appeared strikingly life-like, she was actually a child-size doll—a mechanical contrivance with skin of model wax in a strategically tailored outfit. She was not what many in the audience would surely presume her to be: either a real girl or a costumed doll shell with a real girl hiding inside.

Returning to the stage, Maskelyne would place Zoe atop a stand made of clear glass—not the solid-wood pedestal shown in the promotional photographs—to demonstrate that no cables, pulleys, or other means of machinery were being employed to control her. He would invite a few volunteers to observe the proceedings at close distance, and he would then address Zoe, asking her to draw the likeness of "any celebrity" suggested by the audience. People would call out names—*Charles Dickens! Charles Darwin!*—and Zoe would proceed to draw a portrait of one of them. A poster announcing Zoe's upcoming appearances at the Egyptian Hall in Piccadilly showed an array of ten drawings under the heading "Zoe's Sketches": Dickens, Darwin accompanied by a grimacing monkey, and seven other of the esteemed and/or powerful white men who would qualify as celebrities in nineteenth-century England, including the ever-popular Chancellor Gladstone and Lord Beaconsfield. One woman: the Queen.

A small handful of surviving samples of Zoe's art shows a modest relationship to the meticulous and richly detailed renderings advertised as her sketches in the poster. The drawings she produced with the crayon in her hand are plain and spare, and have a stencilish quality that gives them a feeling of machine production. Audiences and writers covering Zoe's performances were astounded nonetheless by the phenomenon of a humanoid mechanism engaged in art-making without evident human intervention—nor with any suggestion of human agency in the creative process creatively disguised. As a reporter for the *New York Times* wrote, filing from London, "In some mysterious manner she imitates the motions of an artist's arm with facility, holds her crayon in professional form, strikes a true line across her easel in a masterly manner, promptly carries her hand from one point of the drawing board to another, carefully, yet instantaneously, raises her pencil from the paper and transfers it to another spot, returns to add touches and insert omissions, and finally lowers her arm and hand when the sketch is completed."

Dazzled by the mystery in her manner, the *Times* writer ascribed a kind of intelligence to Zoe, noting wryly, "She executes anything she has a mind to."

For Maskelyne, Zoe represented significant progress in an ongoing project to merge science and art, enlightenment and wonderment, through performances that laid claim to presenting both factual reality and fantastical entertainment. Born in 1839 to a family of farmers and rural tradespeople, Maskelyne would lead people to think he was descended from a once-famous scientist with whom he shared two names. Nevil Maskelyne, Astronomer Royal under King George III, was the first person known to calculate the mass of the Earth, and he developed a system for sailors to navigate at sea by diagramming the position of the moon. A mathematician who applied high-level abstractions to practical purposes, Nevil Maskelyne served well as a source of appropriated heritage for John Nevil Maskelyne, a saddler's son who apprenticed in watchmaking and applied a facility with mechanics to the realm of conjuring.

Like Harry Houdini and other practitioners of theatrical illusions, John Nevil Maskelyne was offended by the rise of Spiritualism, the nineteenth-century religious fad led by self-anointed spirit guides who used stage trickery—tricks of the magician's trade—to dupe followers into believing they were experiencing otherworldly phenomenon. *The spirit world is real*, proclaimed the Spiritualists, *and we can prove it tonight in a meeting hall near you, by bringing the ghost of your late aunt Hildegarde into the room, where she will tap out answers to your questions on the side of our Magic Cabinet.* Maskelyne witnessed a performance by the Davenport Brothers, road-tour Spiritualists from America, and set out to debunk their act by duplicating it and exposing its secrets. His intention was not to disprove the religious principle of the spirit; his goal was to discredit charlatans pretending to make spirits palpable through theatrical artifice, a transgression of professional overstepping.

For help with the necessary carpentry, Maskelyne enlisted a friend, George Alfred Cooke, a cabinetmaker with whom he played music in an amateur band. Maskelyne and Cooke replicated the Davenports' effects, including their ability to summon the uncanny sound of musical instruments being played with no musician in sight, and their demonstration was so successful that the two of them decided to work up a full act of illusions, doing real magic: honest deception of the willingly deceived.

According to Maskelyne, the notion to make humanoid automations such as Zoe the centerpiece of Maskelyne and Cooke's act grew from his memories of a childhood visit to the Great Exhibition of 1851 (officially the Great Exhibition of Works of Industry of All Nations, meaning Great Britain, its dozens of colonies at the time, and forty-four "Foreign States"). Housed in the Crystal Palace, a monumental, 840,000-square-foot iron-and-glass greenhouse built for the occasion in Hyde Park, the Exhibition held some 13,000 displays, showcasing such wonders as the world's largest diamond, Matthew Brady's daguerreotypes, and the small, finely wrought work of automation that captured Maskelyne's attention: a singing mechanical bullfinch created by the Swiss artist and inventor Henri-Louis Jaquet-Droz, son of the early automata innovator Pierre Jaquet-Droz, in the late eighteenth century. As Maskelyne would recall in a reminiscence published in the *Strand Magazine*, "The delight of seeing the miniature bird emerge from its hiding-place and sing with all the life and movement of Nature's handiwork was simply indescribable, and I am firmly convinced it was that which first aroused in me a taste for all that is fine and delicate in mechanism."

We can take this testimony as the vivid image-conjuring of a skilled conjurer and discount it for the effects of time, since Maskelyne, at seventy-one, was recalling an event he witnessed when he was eleven. Did the Jaquet-Droz bullfinch, however deftly constructed, truly display *all the life and movement* of a living, singing creature? One answer (among others we'll come to later) tells us something about the need for human beings to simulate the experience of being alive through machinery: We want to believe it can be done. Why else would humans try to do it for literally thousands of years?

There are descriptive accounts of inventive human automations dating to the third century BCE in China, where an emperor of the Han dynasty was said to have had a mechanical orchestra built for his entertainment. There are surviving examples of wooden statuary of human figures with movable legs and arms made in ancient Egypt: sculptures constructed to mimic human activities like plowing soil and kneading dough. Sumatran statuettes with swinging arms and open palms could beat on drums. Dozens and dozens of such creations have been documented and, in some cases, preserved in museums and private collections around the globe. Clearly, the human impulse to represent living things in art carries with it an urge to make art imitate the act of living in a literal, physical way, through mechanical movement. That so much of such art evokes the process of art-making itself—playing music, drawing, writing—speaks to our belief that the ability to make art is one of the attributes that makes us human. To create representations of people able to make art is to become a representation of something greater than human. It's playing God.

The work of Pierre and Henri-Louis Jaquet-Droz, made in collaboration with their protégé Jean-Frédéric Leschot in late-Enlightenment Geneva and Paris, embodies the ideals of verisimilitude and technical precision in mechanical automation. In addition to making the small bullfinch shown at the Great Exhibition, the Jaquet-Droz partners built at least three larger and far more elaborate humanoid constructions that John Nevil Maskelyne eventually learned about: an artist, a musician, and a writer. Pierre and Henri-Louis Jaquet-Droz, well known as makers of high-quality clocks and watches, used the automations as promotional tools, taking them from city to city around Europe to impress potential clients among the wealthy elite. The works must have been impressive in the age before electric motors, since they continue to impress in modern-day

YouTube videos shot at the Musée d'Art et d'Histoire in Neuchatel, Switzerland, the city of Jaquet-Droz's birth, where they have been on display for more than a hundred years.

Each of the three surviving figures is a meticulously detailed, exquisitely attired little person, under three feet in height. Their movements are programmed to suggest the working of human minds, rather than cams and levers. The musician, a young woman sitting at a small, three-octave pipe organ, balances her torso in her seat and follows her fingers with her eyes. Her chest puffs in and out, as if she were breathing, and she makes music by actually pressing the organ keys with her mechanical fingers. The writer, a fleshy boy in a dressing gown seated at a tiny mahogany desk, dips his pen into an inkwell, shakes it a couple of times, moves his hand to the top of the page, pauses as if in thought, and dips his pen back into the inkwell before starting to write in earnest. Every movement is methodically plotted—not just like clockwork, but through actual clockwork—for the purpose of appearing natural and spontaneous. Even so, the tactically humanlike touches in the Jaquet-Droz figures' movements are too abrupt and stiff to come across as truly organic, and the fact that they are precisely identical every time the figures are operated squashes any illusion of spontaneity.

Maskelyne brought the illusion of cognition and feeling, human intelligence, to artificial humanity. The *New York Times* writer in London crystallized the distinction between Jaquet-Droz's marvels of gearwork and Maskelyne's wonders of stagecraft. "'Zoe' bears no resemblance whatever to the celebrated androids of Le Droz [sic], the Swiss mechanician" or others who followed, employing his methods. "They were ingenious pieces of clock-work and nothing more. . . . The work in their case was limited; Zoe . . . can do anything within the power of her secret operator."

Of course, an unseen accomplice had to have been responsible for the presence of cognition and feeling that distinguished Zoe and three more works of automata that Maskelyne produced. Theirs was a vicarious intelligence, supplied by Maskelyne's partner Cooke, working underneath the stage. By 1879, the Maskelyne and Cooke show at Egyptian Hall was featuring four automations, all promoted as being fully independent machines, acting under their own agency, and all reliant on Cooke's handiwork under the floorboards. In addition to Zoe, there was Psycho, a dark-skinned mystery man in a turban and mustache even grander than Maskelyne's, who did mathematical calculations and played the card game whist with volunteers from the audience. The first of Maskelyne's automations to be completed, Psycho bore a conspicuous resemblance to an earlier automation by a Hungarian inventor, Wolfgang von Kempelen, called the Turk: another mechanical man in a turban with skin painted dark to imbue the machine with a racist aura of devilish mystery. The Turk played chess with members of the audience, his movements secretly controlled by a diminutive real man hidden in the cabinetry the automation sat on.

Maskelyne's Psycho was primarily the invention of a farmer and part-time tinkerer named John Algernon Clark, who worked out the secret to hiding Cooke and providing the illusion that all the Maskelyne automata were acting independently. (Maskelyne purchased the design from Clark but took credit for the invention in articles and books published during his lifetime.) The clear-glass stands that supported the figures were actually conduits for Cooke to blow puffs of air into the gearing under their costumes. By stopping and starting, increasing and decreasing the air pressure, Cooke could control the figures' movements. To the audience, the automata appeared to be acting under their own will.

In a typical review of Psycho at work, a writer for *The Times* of London described being struck by the "mysterious power of intelligence," the "secret intelligent force" at work in Psycho's performance. "It is perfectly isolated from any connection—mechanical, electrical, magnetical or otherwise conceivable—with any operator at a distance; and yet, nevertheless, it plays the game of Whist with no little skill. . . . Psycho is perfectly self-acting."

Maskelyne filled out his troupe of mechanical performers with a pair of automated musicians seated in chairs like symphony musicians. Both were reduced-size men in formal wear with a slightly unsettling resemblance to their creator: Fanfare, a cornet player, and Labial, who played the euphonium, a brass instrument similar to a baritone horn. Reviews of their performances describe them playing a selection of pieces accompanied by a pianist from the show's orchestra, as well as duets with Maskelyne and trios with Maskelyne and Cooke, both on trumpet. (The repertoire included once-familiar trifles of Victoriana such as "Hearts and Homes" and "Hark, the Merry Elves.") The participation of Cooke in the group means that another confederate was helping from beneath the stage, and the fact that most of the music was ensemble performance suggests that Maskelyne and Cooke were probably covering for their mechanical band-mates' shortcomings. Nonetheless, the reviews were good.

"The notes are not quite so pure as those of the best human players, but they are very good indeed, and the low notes are particularly satisfactory," wrote a critic for *The Times*.

In a letter to the journal *Musical World*, a euphonium player named A. J. Phasey was even more appreciative. "Labial's observance of light and shade filled me with wonder," he wrote. "The lip action is as perfect as with a human being; and, in his strict attention to the nuances, he was very far superior to many professional artists. . . . I sat close to the stage, to notice whether the fingering was correct or not. I found, to my astonishment, that the passages were not only correctly fingered, but that Labial introduced slurs and appoggiaturas equally as well as any professor of the instrument."

Maskelyne and Cooke billed their act as a show of "Modern Miracles," an alignment of technical innovation and the impossible. They continued to include set pieces debunking seances and Spiritualism, along with conventional stage magic—disappearing, levitation, and a routine in which Maskelyne cut off Cooke's head—as well as the automata that became their signature and most popular attraction. "Whether Messrs. Maskelyne and Cooke successfully expose what they assume to be the humbug of Spiritualism or not, they have clearly demonstrated their capabilities of producing many of the most popular phenomena attributed by spiritualists to supernatural agency," pointed out a writer for the *Observer*. " 'Psycho' still remains an unexplained mystery."

When they first started playing in Egyptian Hall, in 1873, Maskelyne and Cooke had been relegated to a small space on the second floor that seated fewer than a hundred people. It was far from a prestigious booking at a place still associated with the tawdry displays of pseudo-science, colonialist exploitation, humbug, and near fraud that had defined it as a halfpenny "museum," an across-the-pond cousin to the seedy dime shows of titillating exotica and trumped-up "freaks" along the Bowery in New York around the same time. The façade of the building was designed as a hodgepodge of columns, nonsense symbols, and relief effects, complete with statues of a voluptuous semi-nude woman in a Romanesque headdress and a man naked but for a hefty, bulging shield at his crotch: a hall purely Egyptian in the imperial delirium of the nineteenth-century imagination. Prior to Maskelyne and Cooke, the venue had presented such attractions as Siamese twins and the Living Skeleton, a Frenchman afflicted with a tapeworm that prevented his body from metabolizing nourishment.

In the year they introduced automations in their act, Maskelyne and Cooke moved from the upstairs room to the larger main space in Egyptian Hall, and Maskelyne would stay there for thirty-three years, carrying on with other magicians after Cooke fell ill and left the act—the "longest run of any entertainment in the world," Maskelyne would claim. Promoted as "the Home of Wonder," the Egyptian Hall became a full-time showcase for Maskelyne and his automata, performing both in daytime and at night. The act was virtually synonymous with Piccadilly as the district came to dominate popular entertainment in London, and four of the stars were not living things.

A WAVE OF MECHANICAL-PEOPLE ACTS SURFACED IN GREAT BRITAIN AND ITS satellites, in western Europe, and throughout America in the wake of Maskelyne's automata: Zutka the Mysterious, Weston the Walking Auto-mation, Adam Ironsides, Phroso the Mechanical Doll, Steam Man, Ali the Wondrous Electrical Automation, Moto-Phoso, Fontinelle the Auto-Man, Enigmarelle, Perew the Electric Man, and Motogirl, the last of which brought unique dimensions of skill and creativity to her work.

Some of the earliest surviving documents of Motogirl's career place her in Paris in 1903, where, according to an announcement in the French newspaper *Le Gaulois*, she was scheduled to perform at the Olympia Theater. A lovely full-color poster from this period identifies her as "La Motogirl, Célèbre Poupée Electrique" and shows her in a magnificently fussy, over-ruffled green-and-rose dress with a high laced collar, which seems a creation of the same fashion designer Zoe had used in

London. Motogirl is standing upright, pronouncedly upright, apparently under her own power, and has a violin propped against her left shoulder, although the way the fingers in her white gloves are positioned—pressed tightly together and pointing straight out from her sleeves, like five cigarettes poking out from a pack—would make playing the instrument a challenge. A pair of electrical wires dangling from under her dress are connected to the sides of polished black Mary Jane shoes with bows in the front.

A black-and-white photo of Motogirl on a French postcard shows her in the same outfit, sitting stiffly on a high wooden pedestal—like Zoe, again. Her arms are bent in 45-degree angles at the elbows, and two more electrical wires can be seen leading from inside her sleeves to cup attachments on the tips of her middle fingers. In a publicity photo probably from this time, she is once more standing rigidly, arms pressed flat and straight at her sides; she stares ahead in a glassy blank gaze as a female assistant on her right side hands her violin to a man with a mustache in formal wear on her left. Motogirl appears not to be paying attention to what the other two are doing and, in fact, shows no sign of the capacity to do anything, at least not without those wires turned on.

Her act carried distinct echoes of Maskelyne and Zoe at Egyptian Hall, beginning with a mustachioed father figure carrying her onto the stage—in this instance, inside a basket. The man, an American named Frederic Melville, would open the lid to reveal a female doll the size of a young adult folded inside. He would unpack her, snapping her joints into position, one by one, until she was fully standing on the stage, and Melville would introduce what he described as "a wonderful mechanical contrivance made to move and act like a human being by means of a powerful battery."

Like Maskelyne, once more, Melville would carry Motogirl down the aisles, welcoming theatergoers to inspect her, with appropriate care. "I will bring the figure into the audience," Melville would say, "but I must beg of you not to interfere with its delicate mechanism." When someone asked if he could jab the automaton in the eye with his finger, Melville consented matter-of-factly, under a condition. "Certainly," he said, "but as each eye is very delicately made and cost me twenty-five dollars, I shall require the deposit of that sum before you make the experiment." The skeptic demurred, or so said Melville when he recounted the occasion to a writer for the *Strand Magazine*, attesting less to the quality of Motogirl's construction than to that of Melville's improvisational ingenuity.

Inserting a key into a mechanism partly visible in her back, Melville would crank Motogirl into operation, and she would proceed to engage in a small variety of actions that a mechanical device might be expected to accomplish. Melville would press her gloved hands together, and a light would flash like a spark. She would walk stiffly, straight ahead, until she was about to step off the edge of the stage and Melville would catch her, just in time. She would comply unresponsively as Melville bent her limbs and folded her into unnatural, angular positions. Though she was promoted as a "Violin-Playing Robot" and depicted more than once with an instrument in promotional art, no descriptions of her music-making have survived. Touring America in the Keith vaudeville circuit, as Motogirl and Melville did after appearing in Paris, they would have been on stage for ten to twelve minutes of each variety show, and the whole act was simultaneously a presentation of evidence that Motogirl was the "mechanical marvel" she was promoted to be, and a challenge to that possibility.

Melville knew how to stir the public's curiosity about Motogirl's nature, feeding journalists colorful stories to print in advance of Motogirl's appearance in their town. She had been subjected to examination by a team of eye specialists in Boston, "and all the scientists marveled" at the fact that her gaze had been fixed, unchanging—not a blink—for the duration of an hour. She had been inspected and approved as baggage by a customs examiner in Europe. She had been submitted as evidence in a copyright case in Germany and lay motionless on a table during the trial. "It seems almost impossible to believe that for over an hour any human being could undergo such a strain," reported the New Haven *Journal and Courier*—in what may have been one of the first accounts of German copyright proceedings sourced by a vaudeville performer for a Connecticut newspaper.

Descriptions of the Motogirl act would invariably fix on the uncanny quality of her appearance. "She is a very dainty, wiry-looking creature, with all the mechanical charm of a French doll of the latest improved mechanism," reported the *Washington Post*, with the evocation of wires neatly doing double duty. "She is so absurdly natural, and at the same time so artificial, that the spectator, when absolutely confident that she is one thing, finds himself taken unawares in discovering that she is another." Which of the two she was, machinery in the guise of a human or vice versa, the papers left for audiences of the next show to decide for themselves.

Under the porcelain-white makeup, the ruffled smock, and the electrical wires carrying no electricity, Motogirl was a small-framed American actor named Doris Chertney. In a profile published in the *Strand Magazine*, presumably unseen by most Americans attending shows in the Keith circuit, Chertney was reported to have been born to "well-to-do parents, smart society people living within a stone's throw of Central Park" in New York. Her parents died, according to this story, and Chertney was adopted by Melville and his wife, who were living in Havana. Chertney showed a natural ability to sit still and had a high tolerance for physical discomfort, so she and Melville invented Motogirl and took to the road. "I never feel pain. I hardly know what it means," Chertney told the *Strand* writer, "and I never drink tea or coffee, so I have no nerves."

The act was a hit, celebrated in the American press as "the latest vaudeville sensation" and "one of the best shows offered at Keith's in two years." As the *Detroit Free Press* reported, "This 'what is it?' makes 'its' appearances attired in female costume and, notwithstanding that 'it' walks and moves and even talks, no one yet has been willing to wager that 'it' possesses life." Not *she*: "it."

The *New York Times*, in a piece on Motogirl, captured the debate the act was stimulating in the first years of the twentieth century:

"Is it alive?"

"Oh, of course not."

"Well, it certainly does look natural."

"Why, how stupid you are. It don't look a bit natural."

Those are the fragments of opposing opinion as heard in one of the boxes yesterday. Doubtless the same thing is repeated at every performance.

Motogirl, an act with a woman playing a machine playing a woman, tapped a public fascination with the relationship between machinery and humanity that had turned inside-out since the days of John Nevil Maskelyne and his Piccadilly automata. The question given dramatic form on stage was no longer *Could a machine perform like a human being?* but *Could a human being perform like a machine?*

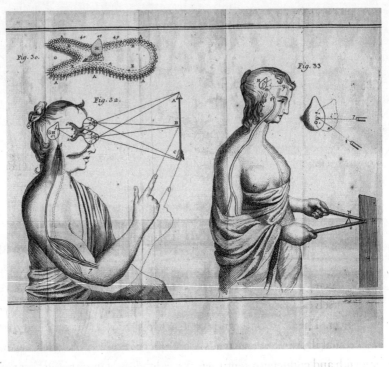

2

THINKING MACHINES

I t's a general law of cultural change that people come to conceive of themselves in the image of the instruments of their aspirations to power. The period we know as the Enlightenment, so named for the intellectual illumination that advanced science and technology in the seventeenth and eighteenth centuries, carried the West into the Machine Age. Pervasive uses of new kinds of machinery transformed the way the Western world worked, in turn changing the way its human occupants understood that world and their place in it. While Jaquet-Droz was inventing machines that looked and moved and made music like real people, actual real people were thinking of themselves as machines.

It is important to remember that all the machinery of this period was not as benign as organ-playing dolls. Cannons and rifles were also machines, and so were the seafaring vessels that brought conquerors to ravage societies and carry off the resources they plundered. So, too, were the cotton gins and milling machines that processed the crops produced with slave labor on American plantations, fueling unprecedented economic growth in Great

Britain. And so were the turbines for water power and steam power that facilitated industrial expansion and the rise of mass production and consumerism. Machines came to dominate Western life in innumerable ways, for better and for worse, doing things only human beings had done before and, increasingly, things humans could not do, at least not on the scale nor at the speed machines made possible and seemingly inevitable. That machines were able not only to emulate human abilities but surpass them in some ways made them all the more alluring as objects of human fascination.

In European thought, the conception of a human being as something akin to a mechanism had roots traceable to ancient Greece and the Pythagoreans, who looked at the body as a whole made up of parts working in synchronous harmony. The Enlightenment philosopher René Descartes advanced this mode of thinking, articulating a spirited argument for the machine as a model for the human body—and the world that bodies occupied. Of course, since he was Descartes, his ideas were Cartesian, ascribing only half of what he saw as constituting humanity, the physical, to the body—with the other half, the non-physical realm of thinking and feeling, attributed to the mind. Not the brain, but the mind. In truth, fifty-fifty is surely not the proportion Descartes was thinking of when he talked of human duality, since he saw the physical as less important to our nature as creatures with a special relationship to God. To Descartes, the body, being a kind of machine, could be understood as part of the vast machinery of the world. But thought and feeling were, in his view, what distinguish us from plants, animals, and the water-powered automated figures in fountains that fascinated him. For Descartes, conceiving of the human body as a machine was a way of reducing it in order to idealize the extra-physical: humanity's link to the spirit, or *spiritus*, the essence of humankind's claim to eternity and the root of creative inspiration.

Fascinated by hydraulic automata, Descartes was a self-taught engineer said to have constructed miniature humanlike figures with moving parts. He was also an ardent student of biology who spent years dissecting animals—and also human cadavers—before writing his *Treatise on Man*, in which he laid out his ideas about the body as a machine. He had seen and touched and studied the interconnecting parts of the human anatomy and even wrote a handbook on the subject, *The Description of the Human*

Body, published posthumously. (See illustration from 1729 edition, page 14.) Like others of his time, Descartes conceived of the heart, not the brain, as the central controlling mechanism for human function—a conception that would be held dear to pop songwriters well into the twenty-first century.

"I suppose the body to be nothing but a statue or machine made of earth," Descartes wrote.

> All the functions I have ascribed to this machine—such as the diges-tion of food, the beating of the heart and arteries, the nourishment and growth of limbs, respiration, waking and sleeping, the reception of the external sense organs of light, sounds, smells, tastes, heat and other such qualities, the imprinting of the ideas of these qualities in the organ of the "common" sense and the imagination, the retention or stamping of these ideas in the memory, the internal movements of the appetites and passions, and finally the external movements of all the limbs. . . . I should like you to consider that these functions follow from the mere arrangement of the machine's organs every bit as naturally as the movements of a clock or other automaton follow from the arrangements of its counter-weights and wheels. . . .

With no place for consciousness in his conception of the human body, Descartes established the model for countless future images of mindless, emotionless vessels that might look like people and perhaps move mechan-ical limbs and even play "Hark, the Merry Elves" on the cornet, but are incapable of thinking or experiencing emotion. Pinocchio before Geppetto wished upon a star, the Scarecrow in *The Wizard of Oz*: they would not be just a nuffin', their heads all full of stuffin', if they only had a brain. They might share their mechanical nature with humans but are missing an ele-ment essential to being fully human, according to Descartes: they don't think. Therefore, they aren't.

Animals and other species of nonhuman creatures were tantamount to automata in Descartes's eyes: organic machines without the capacity for thought. He made the comparison explicit in a letter to the intellectual friar Marin Mersenne: "Suppose that we were equally used to seeing automa-tons which perfectly imitated every one of our actions that it is possible for automatons to imitate; suppose, further, that in spite of this we never took them for anything more than automatons; in this case we should be in no doubt that all the animals which lack reason were automatons too."

The English philosopher Thomas Hobbes liked this view of the body as a machine. "Life is but a motion of limbs," wrote Hobbes in his defining work, *Leviathan*. "For what is the heart, but a spring; and the nerves, but so many strings; and the joints, but so many wheels, giving motion to the whole body, such as was intended by the Artificer." He liked this idea so much, in fact, that he expanded it into a broad mechanical philosophy of everything, explaining politics, business, and every other aspect of Western life though the analog of machinery.

The radical idea that a kind of machinery could be capable of thought—that there could be such a thing as a thinking machine, unrelated to the computer company that trademarked the phrase in the late twentieth century—could be said to have grown from Descartes's ideas, too, but in defiance of them. The French physician and philosopher Julien Offray de La Mettrie took Descartes's conception of the body as a machine and carried it forward, leaving the Cartesian part behind. Finding himself stricken with a fever, La Mettrie noticed that his own mental processes were affected by changes in his heart rate, and he deduced something anathema to the prevailing medical, philosophical, and religious thinking of the time: that something in the mechanics of the body could be the source of human consciousness. Like Descartes, La Mettrie viewed the body as a sophisticated organic machine but came to recognize that the mechanical principles at work in it not only drive the body's physical movements but underlie the workings of the mind.

In 1748, La Mettrie wrote *L'Homme machine* (*Man a Machine* in early translation, *Machine Man* later), a treatise that scandalized the Enlightenment intelligentsia by casting a bit too much light on humankind's inner life and finding nothing magical or mystical there. The book was published anonymously but widely known to be La Mettrie's work, and it was held in contempt for its bluntly atheistic, materialist view of human beings as advanced animals and all animal life as organic machinery. Writing colorfully, zealously, La Mettrie made daring statements with palpable glee. "The human body is a machine which winds its own springs. It is the living image of perpetual movement," he wrote, applying the most sophisticated mechanical technology of the day, clockwork, as a metaphor for the body. "Everything depends on the way our machine is running."

Like an earlier book of his, *Histoire naturelle de l'âme* (*Natural History of the Soul*), La Mettrie's *L'Homme machine* was a slyly audacious refutation of the existence of the soul. "The soul is merely a vain term of which we have no idea and which a good mind should use only to refer to that part of us which thinks," La Mettrie wrote.

Focusing his attention on the function of the brain, La Mettrie went so far as to ground moral character there, rather than in a metaphysical essence. He described the brain as a "subtle and marvelous force . . . (and) the source of all our feelings, of all our pleasures, of all our passions, and of all our thoughts: for the brain has its muscles for thinking, as the legs have muscles for walking. I wish to speak of this impetuous principle that Hippocrates calls *enormon* [soul]. This principle exists and has its seat in the brain at the origin of the nerves, by which it exercises its control over all the rest of the body."

In a leap of the intellectual imagination that had to have perplexed all but the most enlightened minds of his time, La Mettrie anticipated neural activity, asserting, "I believe that thought is so little incompatible with organized matter, that it seems to be one of its properties, like electricity."

Scorned by French society, La Mettrie moved to the Netherlands and, finally, to Berlin, where he wrote the book that further alienated Enlightenment thinkers, *Discours sur le bonheur* (*Discourse on Happiness*). A critique of regret or contrition, the book sought to demonstrate how personal beliefs and behaviors are shaped by enculturation and are not matters of inherent nature or divine influence. In a universe with no God and no souls for human beings to preserve, La Mettrie argued, the purpose of life appears to be the pursuit of sheer pleasure through the gratifications of good food, drink, and physical satisfactions that keep the human machinery humming effortlessly. The book was an entreaty to party like it's 1799.

La Mettrie died at age forty-one after gorging on pheasant pâté with truffles at a feast, apparently contracting a gastric infection in the course of living up to his convictions about indulgence. He was little noted by early writers on the Enlightenment, erased by a de facto decree of *damnatio memoriae* over the anti-religious and pro-hedonism ideas he espoused with hardy fervor. While some of his boldest ideas would not be widely accepted for more than a century after his death, the conception of human beings as machines that he took up after Descartes would have a long, robust life; and his particular notion that an entity conceivable as a machine could be capable of thinking—through the material functions of its machinery—would resurface with special resonance in the late twentieth century.

If all is machinery and thought its defining product, as philosophers after Descartes proposed, God himself (to use the divine gendering of the time) can be thought of as the ultimate machine maker, the master tinkerer and master thinker. David Hume, the Scottish empiricist of the Late Enlightenment, gave voice to this argument in his *Dialogues Concerning Natural Religion*, published in 1779, three years after his death. In a debate between his characters Cleanthes and Philo, Hume has the former proposing that God can be best thought of as a parallel to thought itself. By that thinking, humanity's interest in replicating thought in machinery is not merely playing God but something more audacious: an attempt to replicate God. "Look 'round the world: contemplate the whole and every part of it: You will find it to be nothing but one great machine, subdivided into an infinite number of lesser machines, which again admit of subdivisions, to a degree beyond what human senses and faculties can trace and explain," Cleanthes argues in Hume's text.

> All these various machines, and even their most minute parts, are adjusted to each other with an accuracy, which ravishes into admiration all men, who have ever contemplated them. The curious adapting of means to ends, throughout all nature, resembles exactly, thought it much exceeds, the productions of human contrivance; of human designs, thought, wisdom, and intelligence. Since therefore the effects resemble each other, we are led to infer, by all the rules of analogy, that the causes also resemble; and that the Author of Nature is somewhat similar to the mind of man.

For more than four hundred years now, the machine has served as a malleable analog for human life and the world humans occupy, changing as the character of machinery has mutated from simple two-piece mechanisms to clocks to trains and cars onward to the sphere of electronics and digital devices. The way we employ the human–machine analogy has also changed at least as much as the way we use machines themselves. To say "Man is a machine" in Descartes's time—when clockwork was intricate but comprehensible, and biology was steeped in mystery and conjecture— was to diminish the body, relative to the soul. By the end of the Industrial Age at the turn of the twentieth century, the phrase would open up to a wide variety of meanings and move from the realm of philosophical rumination to the center of the cultural conversation through the popular press of the day.

Journalists relished the handy plasticity of the human–machine parallel: Reporting on a baseball game between the Baltimore Orioles and the Boston Red Sox in 1891, a sportswriter noted that the "infield worked like a machine." *Human–machine as organization of parts working in synchronization.*

In an article of diet advice published in 1903, a health journalist alerted readers, "The human body is like a machine and must be supplied with proper fuel to do its best work for the soul using it. The cells of the body seem almost endowed with intelligence in their work of taking up from the blood the materials needed by different parts of the body . . ." *Human– machine as dependent on energy source but capable of a type* of intelligence.

Another sportswriter, in an account of a boxing match in 1907, wrote of how the prizefighter "Indian Joe" Gregg "kept working like a machine in the clinches and out of them, never giving [his opponent, Steve Kinney] a moment's rest, but bothering him continually with short arm jolts and jabs." *Human–machine as force of relentless speed and consistency of power.*

In an essay on the wonders of human anatomy in 1910, a physician at Johns Hopkins was quoted as gushing, "In very truth, man is a machine more perfect and more wonderful than even the planet on which he dwells. No other is so tenacious of life; no other known mechanism is so efficient." *Human–machine as perfection.*

In a newspaper feature on aging and health in 1912, a physician was quoted describing the body as being "like a machine that has been driven and driven until the rivets and screws are loose." *Human–machine as a set of parts susceptible to wear and tear.*

In 1927, multiple reports of a public debate between Clarence Darrow and Clifton D. Gray, president of Bates College, over the existence of an afterlife recounted Darrow's attention-grabbing comments: "Man lives and works and dies like all other machines. There is nothing to show he is anything else. Any man who is not influenced by hope or fear must believe that. We know what the body is made of. You can buy it all at the drugstore for 95 cents, and I doubt if it's worth that." *Human–machine as cheap commodity.*

MARK TWAIN, BRUISED BY MONEY PROBLEMS AND SHATTERED BY THE DEATHS of his wife and all three of their daughters, fell into depression in his final years. His writing grew darker, bleaker, more cynical, and humorless. In 1906, four years before his death at seventy-four, he wrote a philosophical essay laying out his thoughts on human beings as mindless, hopeless ciphers in a merciless, senseless world. He had it published, at his own expense, in a pamphlet titled "What Is Man?" The phrase was an allusion to Psalms 8:4–6, which, in the King James version of Twain's time, begins with the question "What is man, that thou art mindful of him?"

Twain offered conflicting answers in his text, which was constructed as a Socratic debate between two characters identified as the Old Man (or O.M.) and the Young Man (Y.M.). Twain's late-life point of view pervades the voice of the Old Man, while the Young Man sounds often like the young Twain.

A brief introduction sets the scene:

The Old Man and the Young Man had been conversing. The Old Man had asserted that the human being is merely a machine, and nothing more. The Young Man objected, and asked him to go into particulars and furnish his reasons for his position.

After a bit of casual banter, the Old Man offers up his thesis.

O.M. Man the machine—man the impersonal engine. Whatsoever a man is, is due to his *make*, and to the *influences* brought to bear upon it by his heredities, his habitat, his associations. He is moved, directed, *commanded*, by *exterior* influences—*solely*. He *originates* nothing, not even a thought.

Y.M. Oh, come! Where did I get my opinion that this which you are talking is all foolishness?

O.M. It is a quite natural opinion—indeed an inevitable opinion—but YOU did not create the materials out of which it is formed. They are odds and ends of thoughts, impressions, feelings, fathered unconsciously from a thousand books, a thousand conversations, and from streams of thought and feeling which have flowed down into your heart and brain out of the hearts and brains of centuries of ancestors. *Personally* you did not create even the smallest microscopic fragment of the materials out of which your opinion is made; and personally you cannot claim even the slender merit of *putting the borrowed materials together.* That was done *automatically*—by your mental machinery, in strict accordance with the law of that machinery's construction. And you not only did not make that machinery yourself, but you have *not even any command over it.*

A short time later, the Old Man digs in:

O.M. To me, Man is a machine, made up of many mechanisms, the moral and mental ones acting automatically in accordance with the impulses of an interior Master who is built out of born-temperament and an accumulation of multitudinous outside influences and trainings; a machine whose ONE function is to secure the spiritual contentment of the Master, be his desires good or be they evil; a machine whose Will is absolute and must be obeyed, and always *is* obeyed.

The cover of "What Is Man?" had no author's name. Elsewhere, Twain had made clear that he was proud of the arguments in the book, referring to them in letters as his "gospel." He had been working out the ideas in "What Is Man?" for years, testing them in conversations with friends and in print in a section of *A Connecticut Yankee in King Arthur's Court* (published seventeen years earlier). Still, some unknown consideration kept him from wanting his name on this atypical work of serious, contrarian thought. Was he wary of disappointing fans of the author of the Tom Sawyer books? Did he want the ideas themselves, rather than his considerable fame, to stir interest in the book? Or was the absence of an author's credit an act of meta-affirmation of the book's thesis, that people have the right to claim responsibility for what they might appear to have created by the illusion of free agency? After all, we're just machines.

IN THE MAKING OF MUSIC, THE HUMAN–MACHINE DYNAMIC HAS ALWAYS been more than a philosophical conceit. It has been a practical consideration.

When Henri-Louis Jaquet-Droz introduced his organ-playing automated girl, what made her more lifelike than various music-making automations created before her was the fact that she worked in an overtly mechanical way. In the same manner as a human musician, she played music by applying elemental mechanical laws, using parts of her body in synchronization to manipulate the organ keyboard. A system of cams moved her arms into position, and gears inside her hands transferred force to her fingers, pressing them onto the organ keys. When she hit a key, the note rang out from the organ.

There had been earlier examples of automated dolls that appeared to play mechanical keyboards, though none had actually been playing its instrument. They were fancy music boxes, elaborate variations on souvenir-shop jewelry cases that play a waltz when the lid is opened, but with lovely carved figurines sitting at a nonworking keyboard. Their arms would be geared to toggle back and forth in visual sync but with no other relationship to the sounds the music box produced. Real, living musicians work more like Jaquet-Droz's organ player.

Every human being playing instrumental music has been engaged in processes that could accurately, if tediously, be described as the systematic transfer of energy from their bodies to their instruments through the strategic exploitation of mechanical laws. For untold ages, drummers have used their fingers, the palms and sides and backs of their hands, and the weight of their arms to produce sound in rhythm, aware of the hard work their bodies were doing with no need for technical terms to describe the process. Horn players have drawn in oxygen, expanded and contracted their lungs, and employed their throats and mouths—and, with circular breathing, their cheeks and nasal passages—to force patterns of air pressure into the chambers of their instruments. String musicians have employed fine motor skills to vary the lengths of vibrating wires, manipulating them to produce a range of tones and sonic effects. Musicians knew they were using their muscles, bones, and other body parts to make music long before they learned to frame that knowledge in the language of machinery.

Fig. 110. Method of recording key-movements and vertical finger-movements.

By the turn of the twentieth century, ballooning industrial production and thriving commerce brought unprecedented prosperity to the United States as the population boomed, owing to the demand for bodies to do industrial work. The American public school system expanded and included education in music, art, and the humanities, in demonstration of the New World's aspiration to parity with the Old as a cultured society worthy of its new economic standing. It became a mark of social status for every respectable home in an American town to have a parlor piano, and for every child—or, typically, every girl in the house—to play the piano or another instrument. By the first decade of the century, one in every six households in the U.S. had a piano, with some 2.5 million instruments manufactured and sold in the span of twenty-five years. As the *Wall Street Journal* reported in 1902, "The extent to which pianos have come to be used in this country is amazing."

The teaching of the playing of all those instruments tended to focus (and often fixate) on the mechanics of technique, applying the human–machine philosophy. I should say the *man–machine* school here, because a strain of distinctly male orientation permeated the mechanistic way musicianship came to be characterized in the pedagogy of instrumental instruction. With music seen mainly as a leisure activity of the domestic realm, women's work, the majority of people known to play the piano in America in the early twentieth century were female. Piano instruction, meanwhile, traded in the rhetoric of muscular strength, endurance, and discipline—matters not restricted to any one gender by any means, but generally taken to be the province of manhood at the time.

The dean of this school of thought on the science of musicianship was a pianist, educator, and researcher named Otto Ortmann. Trained at the Peabody Institute, the music conservatory at Johns Hopkins, Ortmann would spend his entire career there, teaching, writing, conducting scientific research on musicianship and musical perception, conducting more research and then still more, and writing more based on his research. Appointed director of the conservatory in 1928, Ortmann set up a laboratory where he led the study of the mechanics of human anatomy as they pertained to playing the piano, along with other projects. He designed and built implements to measure hand movements at the keyboard and the force of individual fingers upon the keys. (See photo, page 25.) He constructed a special pantograph to measure lateral "arm-transfer" of body weight at the keyboard. He used early experimental radiology equipment to photograph the bone structure of the fingers.

Through this and further research, Ortmann wrote an exhaustive study published in 1929, *The Physiological Mechanics of Piano Technique*. Subtitle: *An Experimental Study of the Nature of Muscular Action as Used in Piano Playing and of the Effects Thereof Upon the Piano Key and the Piano Tone.* In 395 pages, the book applied rigorous scrutiny to the human–machine proposition. As Ortmann wrote, "Since the final end of all technical movements at the keyboard is the force-variation, the body becomes a machine, and, like all machines, must obey the laws of mechanical action. Action and reaction equilibrium of forces, dependence of a force upon a mass and acceleration, laws of the lever—all these apply to physiological motion as well as to mechanical motion in general. The fact that physiological motion is more complex, and is under the control of volition, does not introduce any new principle of reaction in the mechanical part of movement."

To quote this book at much length is to open it to likely ridicule it doesn't really deserve. Ortmann's science was less than perfect, and his prose was overstuffed with turgid scientific-ness: "The principle of the parallelogram of forces may be applied to the composition of any number of co-planar forces, by taking each two forces in turn and finding their resultant, which is then combined with one of the remaining forces to find a further partial resultant." Nonetheless, Ortmann was not wrong to recognize that laws of science govern the body. He acknowledged that artistry was also involved, though, and intended primarily to supply tools of understanding for pianists to best use the body for expressive intent. In separate work, he concentrated on the relationship between physical science and psychology in musical perception, aiming to improve both the performance of music and the appreciation of it, through science.

"Otto Ortmann and Rudolf Ganz—I learned their principles. Abby Whiteside knew them and studied directly with them," recalled Sophia Rosoff, a revered piano teacher and director of the Whiteside Foundation, an organization dedicated to carrying on the educational ideals of her mentor, Whiteside, whose work in the mid-twentieth century overlapped with Ortmann's. I interviewed Rosoff in her New York studio and took a piano lesson from her in the first decade of the twenty-first century, when she was in her mid-eighties and still active, teaching the jazz musicians Fred Hersch, Ethan Iverson, Aaron Parks, and others. I came to the interview ignorant of Ortmann and Ganz, and had to look them both up after the session. "I learned from my teacher [Whiteside] that the body contains a set of fulcrums, from the fingers to the hands and up the arms leading to the shoulder. I teach the benefit of playing from the shoulder, and I'll show you what I mean." Rosoff proceeded to direct me to stand—a few feet away from the piano—and do a series of exercises to help me feel the relationship of my shoulders to my hands.

"Ortmann knew more about fulcrums than anybody," Rosoff said. "Science is important. A pianist needs to understand the fingers and the shoulder. But Abby Whiteside taught us something. She taught that music doesn't come from the fingers or the shoulder. It comes from the soul."

I asked how she would teach a pianist to connect the body to the soul. "You have to find your own way," she said and gestured for me to sit back down. "If you can't, don't worry. Your body can do the job."

ABSENT PIANISTS PLAY TO AN ENTHUSIASTIC AUDIENCE AT QUEEN'S HALL.

3

MORE COULD NOT
BE ASKED OF MORTAL
INGENUITY

Are people even necessary? The question is an enduring one for a species that has often seemed oddly determined to prove its own expendability.

As machines became more deeply ingrained in human life, integral to new systems of mass production and commerce in the industrial West, the idea that machinery could replace people—and, in some ways, surpass them in performance—became a truism: if not fully accepted as true, at least recognized as a proposition some people saw as something other people thought of as true. Why shouldn't a machine do the work of a human being, if the human body is essentially a machine, anyway?

In the realm of music, where the importance of physical mechanics had been understood before Otto Ortmann cast it in the tech-manual language of machine-part operations, a wave of inventors and tinkerers tried inverting the human-as-machine principle to see if a machine could make music without the need for a human. In time, the music boxes of the Jaquet-Droz era evolved into progressively elaborate works of engineering and artisanship.

Boxes that made music grew more and more complex and ornate, and larger. And larger. Finely carved hardwood cabinets would be decorated with pearl inlays, sometimes depicting musical instruments against a backdrop of inlaid flowers and fauna. The intent to appeal to monied people of the leisure class was obvious, and the pandering to white privilege could be explicit, as it was in the Swiss boxes designed with a row of small bells geared to be struck by miniature figurines of Black and Asian men in colorful "native" costumes. To wonder today if the owners of such boxes felt unease in the act of manipulating mechanical people of color to repetitive action for their own entertainment would surely be a waste of one's wonder.

By the turn of the twentieth century, equipment to play music automatically was employing multiple technologies—clockwork, springs and gearing, pneumatics, magnetics, and electricity—in wildly complex and imaginative creations. An "orchestra box" the size of a small desk would employ an array of cylinders, chimes, pipes, bells, and drums of various types and sizes to simulate the sound of a chamber orchestra. Replaceable barrels, cylinders, and flat metal discs had the notes for the music encoded in patterns of Braille-style bumps, a resourceful and durable system of pre-digital programming. Through this method of information storage and retrieval, the cylinders would play Mendelssohn's "Priest's March," segments of Wagner's "Lohengrin," popular songs such as Charles K. Harris's "After the Ball," and "God Save the Queen." Yet, the simulation of a violin playing a melodic line would not, could not sound convincingly like a violin; the pitch of the note might be right, but the sonorities were an approximation, a well-intended impersonation of a stringed instrument by a sliver of nickel struck by a tin rod. Beyond being mere approximation, moreover, the note of simulation would carry tones distinctively its own, the timbre of a sliver of nickel. However complex and aesthetically ambitious the equipment, a Symphonium music box would invariably sound much less like a symphony orchestra than what it was: a big, fancy music box.

Furniture stores offered orchestra boxes as parlor adornments with entertainment and educational benefit, showpieces to complement the decorative frou-frou that signified refinement and status in the Victorian home. The Aeolian company promoted its Orchestrelle with ads that pitched the benefits of having a simulated full-scale performance in the home—with content control—at a time when homeowners had not yet become indoctrinated to want such things. "Many people are forced to go through life with never an opportunity of listening to the greatest master-pieces, for even when an orchestra is available they cannot select their own programme, and when a fine symphony or oratorio is heard the effect is fleeting—it has passed before the listener has had time to study or under-stand it," noted an Aeolian ad. "The Aeolian Orchestrelle not only allows everyone to have whatever music they wish, and as often as they wish, but it allows them to play it themselves with full orchestral effects."

Pricey and limited in their appeal, home orchestra boxes such as the Orchestrelle never sold in numbers sufficient to persuade many people that full-scale entertainment could migrate into the home. Meanwhile, larger and more elaborate music boxes were more successful in larger spaces. Exceedingly elaborate music-playing mechanisms, many of them as big as modern-day refrigerators—some even larger, more than twice that size— were produced for use in public places such as beer halls and taverns of the late nineteenth and early twentieth century and, after Prohibition was instituted in 1920, in restaurants and social clubs. Quite a few were coin-operated, taking nickels or dimes and sometimes quarters (a significant sum): jukeboxes of the day. At least one model was connected to smaller remote boxes for selecting songs, like the wall-mounted stations with swinging song menus that roadside diners would have half a century later. Promotional material from their manufacturers to the trade emphasized the practicality of the machines, relative to booking an orchestra of live musicians, paying them for each night's work, and dealing with the vaga-ries of human nature, including the human weakness for the main prod-ucts sold at beer halls. The machines delivered novelty, predictability, and reliability for as long as the madly intricate gearing and beltwork held up.

The novelty value was significant. Much of the appeal in these impressive-looking, overly decorated, clangy, whirring contraptions lay in their looking impressive as they clanged and whirred. Often designed with plate-glass fronts or sliding panels to expose their inner workings, the equipment traded on its technical wondrousness, showing off the com-plexly mechanical reasons it sounded weirdly almost like and yet distinctly unlike an orchestra of living musicians. For some patrons of the places where they were used, these orchestra boxes were an important source of exposure to music and one of the few ways they could hear classical selections, theater songs, and music from social classes other than their own. They were novelties of cultural propagation.

An exemplary specimen of this phenomenon, a grand mechanism with the model name Rex in a line of products called Orchestrions, is intact and still functional, operated daily at 2 p.m. in the Guinness Collection of Mechanical Music and Automata at the Morris Museum in Morristown, New Jersey. Once part of the personal collection of automations owned by a member of the Guinness brewing dynasty, the device was constructed around 1915 by the Popper company of Leipzig, a world center for the manufacture of large-scale automated music equipment. (In the same period, the majority of smaller devices sold in the U.S. were made in Rahway, New Jersey, by the Regina company and promoted as "The Triumph of American Invention" in ads depicting Lady Liberty holding up a glockenspiel.) About ten feet high and six feet wide, Rex contains a kaleidoscopic assortment of implements for the production of musical sounds: horns and organ pipes, bells, a reduced-size xylophone and also a small glockenspiel, various drums, and an arrangement of metal chimes crafted to approximate the sound of a piano, along with objects I couldn't make out through the glass on the front and sides of the oak cabinetry.

On the day I went to see the Guinness collection, in the fall of 2019, a class of middle-school kids was touring the museum, and I got to witness how young specimens of iPhone-era humanity responded to this gargantuan oddity of the Steam Age. The Orchestrion was, after all, nothing like anything in their adolescent lives. Still, it was a kind of prototype for the system of entertainment delivery they were growing up with: a box that housed the most advanced technology of the day for the purpose of translating and storing entertainment content so it could be replayed at will, any time.

A museum guide said a few words, welcoming everyone and acknowledging the gift of the collection from the Guinness heir. He made no mention of beer.

Wearing white gloves, he opened a compartment on the front of the cabinet, and Rex went into action. What sounded very much like the rhythm section of a small jazz band laid out a jaunty dance beat, with a faux piano playing neatly placed filigrees. The artificial piano sounded like a tinny old upright, almost in tune, just a bit flat, and the unpolished, imperfect feeling gave the music a fun quality. A thin simulated horn section joined it as the museum person opened up the bottom half of the cabinet to reveal the labyrinthine machinery at work: hammers tapping in rhythm, rubber belts spinning wooden wheels in synchronization, tiny mallets on hinges flipping to beat drums.

I recognized the song as "Hallelujah" by the composer Vincent You-mans from his musical comedy *Hit the Deck*, a Broadway favorite of the 1920s. An up-tempo number with a catchy, chant-like melody, it was writ-ten to conjure the atmosphere of a revival camp with a tent-show preacher rousing the assembled to a state of euphoria. Though Youmans was white (as were the lyricists for the song, Clifford Grey and Leo Robin), this was music with a lineage in Black traditions—and white applications of Black traditions. When people first heard the Orchestrion play "Hallelujah," in the '20s, most listeners would probably have recognized (or picked up less consciously) that this was something connected to Black life.

The kids in the school group would never know this, of course, although the museum guide, in brief comments after the demonstration, made reference to the fact that music machines such as the Popper Orches-trions were "democratizing" in bringing "music of all kinds" to "work-ing people." This was certainly true: automations brought free (or very low-cost) music to the public sphere at a time of exponential growth in the blue-collar population in American cities; still, it was a partial truth, since people of color were restricted from that sphere. To the extent that the playing of "Hallelujah" in particular constituted democratization, the matter was complicated, with white people of all classes able to hear a machine play a song by a white composer inspired by Black music.

In a Q & A following the formal presentation, one of the students asked, "How does it work?"

Good question. The museum person made a noble effort to explain the principles of pneumatics, through which pumps in the mechanism drove air to make the hammers tap and mallets swing. An incautious mention of how pumps function by "blowing" proved distracting to middle-schoolers.

The mood shifted, and one of the kids barked out, "It's so big! How are you supposed to carry that thing around with you?"

Laughter. To wrap up, the museum rep explained, "Entertainment was still a form of public art" and had not yet moved into the home, let alone into people's pockets.

This was another partial truth and one dispelled by another exhibit elsewhere in the same museum: a player piano from 1923 trademarked as a Duo-Art Pianola. Easily dismissed today as another corny novelty from the era of arm garters and parasols or a museum piece worth a passing glance in a visit to New Jersey but of no special consequence, the player piano is actually an object of transformational importance in the history of American popular culture. A sensationally popular product that made piano music automatically, it changed the very nature of the musical experience in the American home, taking it from one centered on active creativity and individual expression to one all about passive appreciation and curatorship. Along the way, it took a form of Black art—a new school of music invented by artists of color and distinguished by its elementally Black identity—and brought it into the life of people of every race and class. And it did this through a process tantamount to mechanical agency, doing a number of things only a machine could do.

By the first decade of the twentieth century, conventional pianos were firmly established as the centerpiece of American musical culture and family life. Simple or simplified piano arrangements of light classics and popular songs of day—tuneful ditties of Victorian sentiment and valor—were published on sheet music for amateur musicians to play at the family piano. In this time before the rise of radio and record players, performances at the parlor spinet were the means by which most people experienced music most often, and for at least one person in the house, that meant taking piano lessons and sitting at the keyboard, playing the music. That role was generally taken up by (or assigned to) women and girls, who were presumed to have the aesthetic sensitivity suitable for musicianship, as well as the requisite fine motor skills.

Inventors developed varying mechanical techniques to liberate the designated home pianists from keyboard labor and the fleeting domestic glory it accorded, with the females no doubt expected to return to their proper places in the kitchen and the boudoir. The earliest of these devices were accessories called "piano players" that attached to the front of the instruments and did what their name promised: they played the piano with rows of small hammers set up to strike the keys, the instructions for which were encoded in holes punched into spools of rolled-up paper—piano rolls—made to match the written score for the piece to be played. In the late 1890s, these accessories began to be incorporated into special models of instruments, now called player pianos, though most people tended to refer to them by the brand name of the best-selling model of the instruments, the Aeolian company's Pianola. (The trade name became the generic name, the Kleenex or Xerox of musical instruments.)

By 1915, the majority of the dozens of piano manufacturers in the U.S. were offering self-playing models. By 1920, player pianos were outselling traditional instruments. The way most people in America experienced music was no longer through musicianship, by someone in the house performing the music, but through acts of selectivity and commerce and the work of machinery, by buying a piano roll and running a machine to play it. In an ad for the Pianola, the Aeolian company boasted:

> The Pianola is found to be a pleasure-giving instrument for all, both tyro and musician, enables them to play on the piano with absolute correctness and with human feeling. More could not be asked of mortal ingenuity. Less never would have served to place the Pianola where it stands today—the greatest and most widely popular of musical inventions.

The line about "human feeling" was preemptive, a tactic to counteract any doubts about the ability of a self-playing piano to deliver music comparable to the work of a musician skilled at playing with interpretive sensitivity and emotive force. As another Pianola ad stated bluntly, "It is hardly possible to believe until you hear. Piano-playing without an artist! How is it done?" (See illustration for another Pianola ad, page 30.)

The method of doing was ingenious, and it shaped not only the way the piano was played but what it played and how the output was perceived. With no artist in sight, an unexpected art emerged with unforeseen effects.

Player-piano music would typically begin in a conventional way, with a composition created by a human composer. For classical music and work in other styles that involved written scores to be played by pianists, a technician would translate the score to a code readable by the player piano: a set of holes punched into a long scroll of paper. At this early stage in the process, the technology is already influencing the art and not necessarily for the best. The first piano rolls were edited for playing with only fifty-eight or sixty-five notes, though rolls using the full eighty-eight notes of the piano keyboard soon became the standard. Also, significantly, the punched holes could regulate the duration of a note but could not duplicate the myriad variations in touch that would bring feeling—humanity—to the music.

The people tasked with doing the translation from score to piano roll recognized that the player piano, while unable to duplicate the interpretive nuances of human musicianship, had some unique capabilities. A piano roll could be punched to instruct the instrument to play any number of notes at one time, not just the ten possible with the two hands of a person at the keyboard. A roll could also play at rapid speeds that would challenge all but the most technically adept musicians. It was not long before the technicians making piano rolls began functioning creatively, revising the written music to make it sound fuller, denser—less like the playing of a two-handed musician and more like something different: piano-roll music.

An early article in the *Chicago Tribune* about the popularity of player pianos noted matter-of-factly that many piano rolls had "dozens of notes that would not have been written for execution on the piano by human hands for the reason that ten fingers could not encompass them. The music will have been 'orchestrated' beyond the capacity of one pair of hands." Piano-roll manufacturers, a parallel to record companies of a later generation, began labeling rolls for significantly expanded music as "orchestral" editions—symphonic music for solo piano—and started crediting notable roll editors such as M. E. Brown of the U.S. Music Company in Chicago. Her given name was Mary E. Brown, and her trademark was the sly interweaving of musical phrases from multiple works, sometimes layered onto the main melody of the piece. For the piano roll for George Gershwin's hit song "Swanee," for instance, Brown wove in bits of Liszt's "Hungarian Rhapsody" and Dvořák's "Humoresque": audio sampling with a hole punch. She was sampling and mixing, crafting hybrid works of genre-mixing creative curatorship.

Gershwin himself took part in the making of piano rolls for a New York company using a technology that documented what a pianist played at the keyboard, instead of transferring notes from a written score. Through use of this technique, we have piano rolls that convey a sense of Gershwin's piano work, though the rolls are presumed to have been edited and are not quite the equivalent of audio recordings of Gershwin playing. Using a similar method, Igor Stravinsky made half a dozen piano rolls, playing excerpts from his works—*The Firebird, Petrushka, The Rite of Spring, Pulcinella*, and others—at the piano. Stravinsky, an early adapter of new music technologies, even composed a piece especially for the player piano, "Étude pour Pianola," under commission from the Aeolian company in 1917.

The Stravinsky étude, witty and bright-spirited, has a herky-jerky catchiness and urgent, poking rhythms. Taken one way, the piece comes across like a write-around, an exercise in working within limitations: *What accommodations do I need to make in order to sound like a machine?* Taken another way, it feels like a happy experiment in possibilities: *What new things can I do in trying to sound like a machine?*

Critics and classical music professionals tended to see more failings than attributes in the music of player pianos, taking the ways the instruments differed from traditional pianos as detriments. The former concertmaster of the Amsterdam Symphony Orchestra, now teaching music in Los Angeles, found the sound of player-piano music so offensive to his sensibilities that he sued the postcard gallery next to his apartment building for playing piano rolls in the shop. In his complaint, he claimed that the sound of the instrument made him feel as if he had been "struck on the head with a hickory club." The "vulgar" music was so disruptive to his senses that he was "no longer able to detect false from true notes."

A writer for the *Christian Science Monitor* captured the doubts about player pianos well in a piece published in 1911:

Altogether the artificial imitation piano player has ordinary musicians completely eclipsed by its accuracy, tireless energy and its kind, accommodating disposition. And yet the mechanical player gives a colorless, emotionless, hollow picture always the same though it may be taken direct from a Paderewski. It lacks all the charm and beauty of a human interpretation.

It gives them [who can't have a living player play] a sort of photographic idea of how music might sound. . . . Still it is not art. It occupies the same relation to art that a photograph of an oil painting does to the painting itself.

It plays ragtime exceptionally well and entertains a great many people who do not necessarily care for musical art.

These criticisms were widespread: player pianos, being machines strictly regulated by their mechanics, produced music devoid of the shading, the sensitivity, and the interpretive imagination that human musicians could provide. The sound lacked personality and a core value of aesthetic quality: feeling.

How, then, could these instruments play one particular style of music, ragtime, exceptionally well? What did such a statement—slipped in casually, almost as an afterthought—mean to say about player pianos and the ragtime genre?

One point is made plain: ragtime was not to be thought of as musical art. The same proposition appeared in other articles *defending* player pianos and portraying their users as sophisticated aesthetes. As early as 1904, the *San Francisco Chronicle* was describing users of self-playing pianos as "people of culture and refinement, lovers of music, and of good taste in their preferences." Then came a qualifier: "Yet many of them first invested in the automatic piano player for the amusement of hearing ragtime unrestricted."

Something unstated about ragtime appeared to disqualify it from consideration as a musical art in the eyes of cultural mediators in the mainstream press—that is, the white press—and that was not its sound but its color. Ragtime was a form of resoundingly Black expression. Named for its distinctive syncopations in rhythmic phrasing, or "ragged time," the music was highly complex and musically imaginative, constructed through the accretion of individualized thematic variations and distinguished in part by its subtle but essential use of "blue" notes, touches of dissonance employed judiciously to bring emotional tension to the music. Ragtime was challenging to play, requiring exceptional dexterity in both hands, particularly the left, which would plot out intricate repeating patterns while the right hand played literally offbeat (off the beat) melody lines and contrapuntal patterns. Playing this music tasked the skills of the amateur musicians who comprised the bulk of the market for popular music in the sheet-music era.

Ragtime called for all the qualities critics valued in classical music: shading, sensitivity, interpretive imagination, and feeling—and, also, exceptional technical facility and speed, as well as mastery of musical time. As the music became popular, people (including critics) appeared to lose sight of the relevance of humanistic values in the playing of ragtime, associating the music mainly with technical and mechanical values. Player pianos deserve part of the blame.

While ragtime was first established in the form of music scores and sold well as song sheets for home pianists, it took off as a popular phenomenon—more than that, a full-fledged craze—on piano rolls for player pianos. By the mid-1920s, piano rolls were selling at the rate of 11 million copies per year, with some fifty companies in the roll business. More than 1,000 titles on piano rolls were ragtime pieces. Dozens of rags by composers such Eubie Blake, James P. Johnson, and others were released only on rolls and not even published in sheet music.

The packaging for piano rolls rarely depicted the composers in imagery, downplaying or outright hiding their race. Though ragtime was generally understood to be a style of Black music, a white listener could buy a piano roll and have it performed on the piano in the living room without allowing an actual person of color in the house. They could take pleasure in it, reveling in Black creativity, but in detachment. That white listeners could control the faceless, impersonal player-piano machinery at will—running the music too fast for comic effect, making it play over and over and over—added a profoundly troubling dimension to the three-way dynamic of artist, listener, and delivery machinery.

The technologies of piano rolls and player pianos seemed well suited to a music that was technically demanding and called for minutely precise articulation of notes in arabesque patterns. In fact, much of our conception of ragtime as a music about precision and speed is a perception informed by the way the music was adopted for use on player pianos and adapted to exploit their potential, at times producing music impossible for a two-handed human being to play. When we think of ragtime, we hear player pianos in our minds. The machines shaped the way the music is understood.

In the process, player pianos—machines that made music with no one at the keyboard—brought Black music into American homes of all races and classes. Fortunately, it never left.

≈

CONLON NANCARROW, A TWENTIETH-CENTURY COMPOSER FROM TEXARKANA, Arkansas, was a social idealist and musical adventurer intolerant of human shortcomings. When he began composing professionally, in the early 1930s, he found it frustrating to work with musicians who struggled to grasp his unorthodox ideas, especially his radical approach to rhythm, and he started "dreaming of getting rid of the performers."

Nancarrow found hope in writing by the iconoclastic composer and theorist Henry Cowell, who admired Stravinsky's experiments with the player piano and saw further potential in the instrument. Other composers, including Paul Hindemith and George Antheil, were experimenting with player pianos as well, with Hindemith writing a fiery piece, *Toccata for Player Piano*, that sounds not merely like more than one pianist playing at the same time, but like multiple player pianos playing simultaneously. Nancarrow put aside music temporarily to fight for the Lincoln Brigade, the militia led by the Communist Party in support of Spain's Republican Loyalists in the war with Franco. When he returned to the U.S. in the late 1930s, he felt stigmatized by association with Communism and moved to Mexico City, where he would live for the rest of his life. Working alone in semi-voluntary exile, he developed a way to make music that got rid of the performers.

"I was always constrained by players' limitations," Nancarrow said in an interview late in life. "With the player piano, I just did what I wanted to do." Beyond the relatively practical matter of musicians' physical capabilities, Nancarrow was drawn to the player piano's mechanically ethereal spirit. "All those keys going in the middle of it . . . no one playing it," he mused. "It's a special thing."

Adapting the methods of player-piano music editors such as Mary E. Brown, Nancarrow worked out a process for composing directly onto paper rolls. He would sit at a hole-punching machine he had custom-made, figure out the spot on the paper where he wanted a musical note, and squeeze a handle to punch the note out of the paper. The hole-punching process for one piano piece of two or three minutes' length would take Nancarrow several months. Part creative purist, part garage tinkerer, part secular monk, Nancarrow spent every day in his Mexico City studio, submitting to the intensive physicality of a musical form known for requiring no human effort.

Over four decades, Nancarrow created dozens of musical works designed to exploit the extra-human possibilities of the player piano. Though individually brief, each no longer than a pop radio tune, Nancarrow's pieces are rich and dense, miniatures in duration but not in ambition or impact. A Nancarrow composition is a musical hurricane happening during an earthquake at the time of a sunspot shower: an overwhelming assault of ever-changing rhythms, chord patterns, and melodic ideas. Contrapuntal phrases overlap; walls of notes rise, whirl, and smash apart. A great many of Nancarrow's compositions push and pull the lines between formal and informal expression, classical and popular music, the avantgarde and the blues. "Most of my things are something against something else," Nancarrow explained.

After some four decades of neglect, Nancarrow was discovered by the music world in the 1970s and enthusiastically embraced over the decades that followed, thanks in large part to the advocacy of New Music scholars Peter Garland and Kyle Gann. "His music is so utterly original, enjoyable, perfectly constructed, but at the same time emotional. For me, it's the best music of any composer living today," said the composer György Ligeti in 1981. The following year, Nancarrow was honored with a MacArthur "Genius" grant.

In photos accompanying the appreciative articles that began to appear in the mainstream press as well as in music publications, Nancarrow seemed ideally cast for the role of monastic genius: thin and owl-eyed, with flowing white hair and a trim, pointy goatee. Wary of attention, he spoke to the few interviewers who sought him out in whispery, cryptic phrases. "Music is for listening," Nancarrow said, and he bore that out in his unorthodox pieces, which were generally a pleasure, as well as a challenge, to experience. Through produced by utterly mechanical means, they served the human spirit.

The player pianos all the rage in the early twentieth century made plain that some pianos were highly mechanical devices. Yet the fact that all pianos have always been machines—not only the Pianolas churning out ragtime from paper rolls but also the nine-foot concert grand Vladimir Horowitz played in Carnegie Hall—is an obviousness to which most of us would probably not give much consideration when we're enjoying (or, for that matter, disliking) the sound of piano music. If human beings can be thought of as machines in a figurative sense, as Descartes taught, pianos can usefully be thought of as machines in literal terms, because mechanical constructions are what they are. In fact, a great many musical instruments rely upon the interworking of moving components and thus qualify as machinery. The vast majority of instruments most common in Western music—nearly all the pieces in a symphony orchestra other than the timpani and the triangle; almost all of the pieces in a rock band or a jazz group, including the kick bass and the hi-hat in the drum kit; and even most of the pieces in an old-time Appalachian string band, not counting the washboard—have always been machines. The flute has some 120 parts, with hinged keys and springs that move to produce finite variations in sound. The acoustic guitar has an intricate set of moving parts, some (the gears of the tuning pegs) employed for the benefit of others (the strings, whose movement in the form of vibration provides the instrument's sound).

When Woody Guthrie sang his earthy sing-along songs of political dissent in union halls, he was playing a guitar with a hand-lettered message, "THIS MACHINE KILLS FASCISTS," painted on the body. "Guthrie began to think of his acoustic guitar as a machine: a tool, like a lathe, or maybe even a weapon, like a bullet, bomb, or gun," wrote the historian Michael J. Kramer.

Pianists have long been acutely aware of the mechanical character of their instrument, recognizing that its musical potential is tied to its particulars as a mechanism. "The piano is a machine, and a good performer ought to know his instrument as well as a good chauffeur knows his car," Josef Hofmann, the classical pianist and composer, told a reporter for the *New York Times* in 1927. "He ought to understand its mechanical action and possibilities." The pianist can play only what the machinery allows.

Musicians sometimes talk about their instruments in the language of machinery as an act of deference, in respect to the technical demands of the instrument. "The piano is a percussion instrument—no more and no less, but it's a *mechanical* instrument," said the pianist Donald Shirley, who was known for playing spirituals, jazz standards, and classical works with delicate precision. "Do you understand what that *means*? I'll tell you what that means. That means that the piano is a *force* to contend with. It's not like a damn typewriter, where you just tap-tap-tap, and it does what you tell it to. The piano has a will of its own. It's a machine with a *brain*. It may look like it's a big box with strings and whatnot in it. But it's a *machine*, and it will only do what it *wants* to do.

"If you intend to play the piano, you have to learn to work with it, or it will *tear you apart*. Do you understand what I'm saying?"

It may not be easy to understand how musicians such as Donald Shirley could see their instruments in mechanical terms while anthropomorphizing them and projecting aggression upon them. Shirley had no pet name for his Steinway concert grand, the way Willie Nelson called his Martin classical guitar Trigger (after Roy Rogers's trusty stallion) and B.B. King referred to his Gibson electric guitar as Lucille (after a woman two men were fighting over when King almost lost the guitar, as well as his life). But Shirley held his instrument dear and pampered it, running an industrial humidifier near it twenty-four hours a day, and he understood its inner workings like a mechanical engineer.

In the twenty-first century, the expansive use of electronics, digital technology, and artificial intelligence in the making of music and imagery can tempt us to mistake the presence of machinery in the creative process as something radically new and threatening to the authenticity of the work. Yet humans have been working with machines to make music for as long as people have been playing instruments with strings or valves or keys. People and machinery have come together to make all kinds of art for a very, very long time, with the machines exerting influence on the art that's made. Changes in technology over time—what the equipment can do, how it does it, what it in turn allows people to do, and how they do it—can be transformational to the art and the culture it's a part of.

Herbie Hancock, a virtuoso eclecticist, built a legacy of fruitful experimentation with a wide variety of keyboard instruments, and he found that the technical particulars of each one influenced his playing. "I play a little differently on the Clavitar [a guitar-like keyboard instrument strapped over the shoulders] than I do on the E-MU [a polyphonic electronic keyboard] or when I'm playing the Fazioli [the concert grand piano of Hancock's choice]," he told me. "I'm not just talking about the fact that the feeling and technical aspects of music varies from a funk thing with an electric band to a solo piano piece. I'm talking about the character of the instruments themselves. Every instrument I play has its own character. I pay attention to the instruments, and I respond to them, the same way I listen to the drummer or the bass player in my band and respond to them.

"When I was playing with Miles, I started on acoustic piano, and when I switched to electric, it played the tunes differently. The tone and the touch, the sustain, the whole atmosphere, the personality of the music—everything about it—was suddenly different. That was the electric keyboard leading me. I followed the keyboard into electric wonderland."

4

EVEN THE KITCHEN SINK

T he walls were unpainted, just white plaster, and vertical girders in the open space and crossbeams on the ceiling were exposed. For light fixtures, hardware-store tin pails served as lampshades over incandescent bulbs in plain ceramic fixtures. There was no carpeting. The building, a new thirteen-story tower half a block from Carnegie Hall in New York, was not yet finished but was serving, temporarily, as the setting for an art show. At the time, the spring of 1927, the term "industrial" had not been established as a category of chic interior design, though the unadorned, utilitarian atmosphere of industry was precisely what the space conveyed, and its assigned purpose was the essence of intellectual chic. For two weeks beginning on May 16, the ground floor of 119 West 57th Street was home to the Machine Age Exposition, a gallery-style public exhibition of objects never previously assembled in a gallery: industrial materials, machine parts, architectural plans and models, and other artifacts of the intersection of art and machines. More than four hundred exhibits from seven countries were presented on simple wooden stands and hung on the walls: motorboat propellers, radio sets, deep-sea-diving suits, slicing machines, ventilators, valves and gears and machine components, steel cupboards, rifles and tommy guns, automotive and architectural design drawings, and farm tools, along with photos and paintings with industrial themes.

Item number 422, "Steel Frame," and items 423 and 424, "Moving Parts," were on loan from Steinway & Sons, the piano manufacturer. The frame, which in 1927 would have been made of a metal alloy closer to gray iron than steel as we know it today, was the skeletal structure for a specimen of one of the most highly regarded musical instruments made in the United States. There is no surviving record of which two of a Steinway piano's many smaller parts were selected to be displayed; a Model M, the most popular Steinway grand in that period, had more than 12,000 components: flanges, levers, whippens, dampers, lyre rods, shanks, heels, capstans . . . In an exhibition intended mainly to challenge visitors to think of machinery as art, items 422–424 asked them to consider a musical instrument as a type of machine—an object of art that was also a machine for making a different kind of art: music.

The Machine Age Exposition was an audacious public display of ways in which machines had come to shape how artists, intellectuals, laborers, and combinations thereof worked and thought about work in the early twentieth century. Sponsored by a small but fierce journal of ideas, *The Little Review*, the exposition mobilized objects from everyday life and industry, juxtaposed with works of more overt artistry, to demonstrate the aesthetic value and the sociological import, and—why stop there?—the sociopolitical and the socio-religious relevance of machines, machine-made art, and art concerned with machines and their impact. "That the machine is the tutelary symbol of the universal dynamic can be discovered at the Machine Age Exposition . . . instigated by Miss Jane Heap," wrote E. B. White in a "Talk of the Town" item in *The New Yorker*. "Miss Heap believes that art, so long as this is a cock-eyed world, must be viewed from crazy angles."

Haughty sarcasm notwithstanding, White noted correctly that the show presented the machine as a capacious emblem of its time, and it was true that the event's primary organizer had ways of looking at things that were unconventional for her day. Principal editor of *The Little Review*, Jane Heap had been working in close collaboration with the magazine's founder, Margaret Anderson, since the two had met in the magazine's Chicago office eleven years earlier and clicked as compatibly bookish, free-thinking devotees of the arts. Heap had grown up in Topeka, Kansas, in a house next door to a mental institution where her father worked as direc-tor (warden of the insane asylum, in the language of the time). "From the insane I could get everything," she would later write. "They knew every-thing about nothing, and were my authority."

Heap made an arresting impression, invariably dressed in a man's frock coat over a white tuxedo shirt and laconically tied bowtie. She had a strong, long, rectangular face, which she set off with dark hair cut in a pompadour that flopped over her forehead like bangs. Heap never went out without putting on fiery red lipstick. Gifted as a visual artist, Heap had studied at the Art Institute of Chicago and taught art for a few years before connecting with Anderson both professionally and personally. (Heap called Anderson "Martie" or "Mart" and described her in a letter to her dear friend Florence Reynolds as "my blessed antagonistic comple-ment and antithesis.") They moved together from Chicago to New York, set up a little office for *The Little Review* in Greenwich Village, and set up housekeeping a couple of blocks north, in Chelsea. It was Heap who came up with the journal's credo: "To express the emotions of life is to live. To express the life of emotions is to make art"; and it was Heap, the person who had learned the wisdom in nothing from asylum inmates, who decided to print twelve blank pages in the September 1916 issue of *The Little Review* because an insufficient amount of quality writing had been submitted for publication.

The journal nearly folded after publishing an essay by the anarchist Emma Goldman calling for the abolition of religion and private property, prompting several of its financial backers to withdraw their support. With the help of Ezra Pound, who signed on as international editor, the *Review* brought James Joyce's *Ulysses* to American readers, serializing chapters until the U.S. Post Office seized copies and the editors were arrested on obscenity charges. Anderson and Heap went to trial in a closely watched and much debated case study in state oppression of literary speech, and they lost, though they managed to keep *The Little Review* in print. By May 1927, when Heap oversaw the journal's launch of the Machine Age

Exposition, the imprimatur of *The Little Review* was a mark forewarning bold, unexpected, and, for some, unacceptable ideas.

Heap, who rarely published writing of her own, wrote an essay lauding the aesthetics of machinery for the magazine and adapted it as the introduction to the exhibition catalog. "There is a great new race of men in America: the Engineer. He has created a new mechanical world [and] is segregated from men in other activities," Heap wrote.

> The men who hold first rank in the plastic arts today are the men who are organizing and transforming the realities of our age into a dynamic beauty. They do not copy or imitate the Machine, they do not worship the Machine—they recognize it as one of the realities. In fact it is the engineer who has been forced, in his creation, to use most of the forms once used by the artist. . . . The artist must now discover new forms for himself. It is this "plastic-mechanical analogy" which we wish to present.

If the artists she valorized were not guilty of machine worship, Heap herself had an inclination to it. Her thinking on the relationship between machinery and humanity was informed by the teachings of the early-century mystic George Gurdjieff, a self-styled mysterioso—elusive origins, unplaceable accent, shaved head, and gigantic curlicue mustache—who saw human beings as organic machines whose actions and thoughts were strictly prescribed by the laws of nature. He taught that it was only after understanding humanity's nature as machinery that an individual could begin taking steps to achieve a kind of spiritual transcendence, under Gurdjieff's guidance through a process called "the work." Donations welcome. "Every one of you is a rather uninteresting example of an animated automation," he told his acolytes. "You think that a 'soul,' and even a 'spirit,' is necessary to do what you do and live as you live. But perhaps it is enough to have a key for winding up the spring to your mechanism."

Like the dogma of later cults and quasi-spiritual movements—and, for that matter, anti-spiritual movements—Gurdjieff's doctrine was elliptical, semi-contradictory, and alluringly vague. What Heap took away from it, mainly, was a fascination with machinery and its ability to hold humans in sway.

At the same time, Heap was drawn to a contemporaneous but differing conception of machines from another intellectual power center: the Italian artists and writers known as the *Futuristi*, the Futurists, whose colorful and aggressively provocative ideas made them voguish among avantgardists in the era of early modernity. Led by the poet Filippo Tommaso Marinetti, a group of militant aesthetes from Italy published a series of self-defined manifestos for a new day of radical creativity. They held up the machine as a sacred totem of nontraditional ideals: artifice, insensitivity, rashness, ferocity, irreverence, peril, and above all speed. With their strident calls for cultural and political upheaval and their idealization of violence, the Futuristi anticipated the rise of Italian Fascism and played a significant role in enabling it. Heap welcomed articles on the radical aesthetics of Futurism in *The Little Review* and made a Futurist proclamation the centerpiece of the catalog for the Machine Age Exposition. Written by Enrico Prampolini, "The Aesthetic of the Machine and Mechanical Introspection in Art" announced:

> Is not the machine today the most exuberant of the mystery of human creation? Is it not the new mythical deity which weaves the legends and histories of the contemporary human drama? The Machine in its practical and material function comes to have today in human concepts and thoughts the significance of an ideal and spiritual inspiration.
>
> We, today, after having sung and exalted the suggestive inspirational force of the Machine—after having by means of the first plastic works of the new school fixed pure plastic sensations and emotions, see now the outlines of the new aesthetic of The Machine appearing on the horizon like a fly wheel all fiery from Eternal Motion.
>
> We therefore proclaim:
>
> 1. The Machine to be the tutelary symbol of the universal dynamism, potentially embodying in itself the essential elements

of human creation: the discoverer of fresh developments in modern aesthetics.

2. The aesthetic virtues of the machine and the metaphysical meaning of its motions and movements constitute the new font of inspiration for the evolution and development of contemporaneous plastic arts.

3. The plastic exaltation of The Machine and the mechanical elements must not be conceived in their exterior reality, that is in formal representations of the elements which make up The Machine itself, but rather in the plastic mechanical analogy that The Machine suggests to us in connection with various spiritual realities.

4. The stylistic modifications of Mechanical Art arise from The Machine-as-interferential-element.

5. The machine marks the rhythm of human psychology and beats time for our spiritual exaltations. Therefore it is inevitable and consequent to the evolution of the plastic arts of our day.

It was probably to the show's advantage that coverage of the Machine Age Exposition in the general interest press was terse and indifferent, as if newspaper writers were not quite prepared to say much of substance on the subject of machine parts, industrial objects, design plans, and the like offered up as art. A few specialized journals with liberal orientations published serious reviews, and they were sympathetic to a project sponsored by a small journal known to be habitually struggling. "Machines themselves, approached aesthetically, appear to resemble other works of art in that they are not all always beautiful," wrote the architect and critic Herbert Lippmann in *The Arts.* "Evidently sculpture and painting have the edge on machines as exhibition material in that they can look their best in a gallery, but the machine will look better on the job."

The poet Genevieve Taggard reviewed the exposition for *New Masses,* the socialist monthly, and had some tempered praise for the show's previously untested notion to repurpose machines for critical consideration. "It may be that the machine has got to be stopped before we can see it," Taggard wrote. "In that case this show was right in taking from the delicatessen and the machine shop the tools and contrivances we watch with bored eyes and impatient faces, rapping the counter with the change while we wait for sliced ham or re-soled shoes."

Ultimately, however, Taggard saw stopping the machines as a misrepresentation of their nature that failed to convey the dynamism of modernity. "There could have been more guillotineques, nearly noiseless meat-slicers from Dayton, more kitchen cabinets and Crane Valves; *more Machine Age*," Taggard argued. "After the show we went outside into a comparatively better show, the city of New York, mixed up with all the past, to be sure, mixed as all art is in life—but superior to Miss Heap's show in two regards: first, there was more of it and second, it was going."

Four days into the Machine Age Exposition, on May 20, 1927, Charles Lindbergh took off from the Roosevelt airfield in Long Island for the 33.5-hour flight that would earn him a permanent place in every grade-school history book as the first person to fly alone across the Atlantic Ocean. He came home a hero and, more notably, a celebrity: a photogenic white male glorified in the new mass media of newsreels, magazines, and radio as a symbol of the can-do spunk and rugged independence of the American myth. That the literal vehicle of his heroism was one of the great marvels of the machine age, a flying machine, was essential to the narrative of Lindbergh as a new kind of hero. When the Futurists raved about *speed* and Genevieve Taggard gushed about *going*, they were talking about trains zooming down tracks and cars zipping around city streets. With aircraft—even single-engine prop planes like Lindbergh's *Spirit of St. Louis*—machines not only carried people faster, but took them higher and farther. Lindbergh was duly celebrated for the unprecedented achievement of his flight, an act of extraordinary fortitude, discipline, and bravery. Of course, other men and women throughout history had shown extraordinary fortitude, discipline, and bravery. The uniqueness in his story lay not only in what Lindbergh achieved but also in what his flying machine accomplished by carrying a man across the ocean through the air. The man and the machine had worked together in synchronicity—in a sense, in collaboration.

Within weeks of Lindbergh's flight, another young flyer, Amelia Earhart, began a campaign to urge more women to take up flying. "If women would once try flying they would lose their fear of it," she told the *Boston Globe*. "To my mind, there is as much danger on a yacht as there is in an airplane." Within a year, Earhart would set a new record as the first woman to make successful solo flights across the United States in both directions. Four years later, she would duplicate Lindbergh's big flight, becoming the first woman to fly alone across the Atlantic. She wrote a nationally syndicated newspaper column and canvassed the country, lecturing on aviation. She was toasted by President Coolidge at the White House, lauded as Lady Lindy and the Queen of the Air—an aerial hero and celebrity to match Lindbergh in stature as well as good looks. (As writers of the time sometimes noted, the two flyers had an eerie physical resemblance.) Typically photographed in the cockpit of a plane or working on an engine in flight suits she had to design and sew herself because all the standard gear was tailored for men, she embodied a steely and cerebral, un-girly femininity.

When the Museum of Modern Art opened its first major exhibition of mechanical objects, seven years after Jane Heap's Machine Age Exposition, the curators assembled a jury of renowned public figures to review the works on display and select one piece as "the most beautiful object in the exhibition." The jurors were the philosopher John Dewey, who had just published a book of political philosophy, *The Public and Its Problems*, which, among other things, critiqued the cultural obsession with technology; the educator Charles R. Richards, president of Lehigh University and cofounder of the Museum of Science and Industry; and Amelia Earhart. (See photo, page 48.) A fourth person, Frances Perkins, Secretary of Labor, was not able to participate but weighed in after looking over the photos in the exhibition catalog.

The three judges had three different top choices: Dewey, an aluminum propeller for an outboard motor; Richards, a set of large steel balls made for an oversize ball bearing in factory equipment; Earhart, a section of stainless steel spring. Thick and strong, it had soft, warm curves—no hard edges anywhere—and had been polished to the reflective gloss of a mirror. From the angle the photographer used for the catalog photo, the spring piece suggested a copyright symbol in 3-D, rendered with the hyperrealistic, glistening metallic perfection later generations would associate with computer-generated effects. After deliberating at the museum, the three judges decided to go with Earhart's choice.

Earhart said she admired the ball bearing and, indeed, used to keep a ball bearing on her desk, along with other machine parts she found attractive. "As a matter of fact, in designing clothes, I made a buckle for a belt out of that ball bearing because I thought it was a beautiful thing," she said in comments the museum later distributed to the press. "I have on my desk as an ornament an intake valve because it is beautiful and this section of spring I chose first because it is satisfactory from every angle. One doesn't have to know what it is to appreciate its form, and I think that the exhibition as a whole is a great step forward in that I believe people may see beauty in machine [sic] which so often we think of as only crude or lacking individuality, which isn't the case at all for anyone who has eyes to see."

The exhibition Earhart and her fellow jurists surveyed for the benefit of the Museum of Modern Art public relations department was a high-profile New York cultural event called "Machine Art," and it was only harmlessly—and perhaps helpfully—tinged with the aura of prankishness implicit in the presence of John Dewey and Amelia Earhart acting essentially as beauty contest judges. Presented from March 6 to April 30, 1934, the show filled all four floors of the gilded former Barbour mansion on West 53rd Street in Manhattan, the museum's second location in the five years since it had been founded by members of the Rockefeller family and their friends. The site exuded old money and its privileges, while, for eight weeks, its contents conveyed nothing but newness in the form of mechanical devices, mass-produced objects, and materials for modern living.

"Machine Art" was the second newsmaking show within two years to be organized by the museum's founding director, Alfred H. Barr Jr., and his exceedingly ambitious, equally charming, and surpassingly rich young protégé, Philip Johnson. A recent Harvard graduate who had traveled around Europe and emanated cosmopolitan élan, Johnson personally donated the money to start an architecture program at the museum, and he was advanced from a spot on the Junior Advisor Committee to head up the new initiative. Aged twenty-five in 1932, he co-curated the first exhibition of contemporary architecture at MoMA, working in collaboration with the historian Henry-Russell Hitchcock Jr. The show highlighted the work of Walter Gropius, Le Corbusier, and Ludwig Mies van der Rohe, among others (including the American Frank Lloyd Wright) through models and drawings, and played a major role in establishing the aesthetic of their clean, unfussy, geometric designs as a school of modernist art: the International School, as Johnson would call it.

Following up with the "Machine Art" exhibition, Johnson sought to situate the International School in a broader cultural context, selling it as a bigger story by showing the permeating influence of its underlying principles as they were manifest in the machinery and machine-made objects that pervaded and, arguably, defined modernity. The show displayed nearly one thousand objects collected by Johnson and his research committee over six months of visits to the factory showrooms of American Radiator, Carnegie Steel, Corning Glass Works, Electromaster, Hamilton Beach, National Cash Register, York Safe & Lock, and other manufacturers. Some of the pieces they selected were the kinds of things Jane Heap had chosen for the Machine Age Exposition: springs and engine components, propellers, and such. In addition, "Machine Art" included an expansive array of household appliances and kitchenware, scientific instruments, office equipment, and odds and ends. Objects were displayed with austere elegance on black velvet draped over cedar pedestals. Throughout the rooms of the museum, false ceilings and moveable wall panels of stretched muslin in colors of pink, gray, and blue covered up the building's decorative molding and wainscotting. Under Johnson's direction, the rococo interior of a blue-blood temple from the Gilded Age looked suddenly like the inside of an icy Mies van der Rohe box building.

"There are no purely ornamental objects; the useful objects were, however, chosen for their aesthetic quality," Johnson explained in the exhibition catalog. "Some will claim that usefulness is more important than beauty, or that usefulness makes an object beautiful. This Exhibition has been assembled from the point of view that though usefulness is an essential, appearance has at least as great a value."

Contrasting machine-made objects with materials produced by living, breathing artists and craftspeople, Johnson contested the primacy of the human touch, something elemental to the handcrafts movement that had raised the standing of the folk arts during the Depression. Johnson argued, in a radical statement that called for further explication, that the processes machinery employed were demonstrations of a phenomenon historically held up as unique to human art makers: technique. "In spirit machine art and handicraft are diametrically opposed. Handicraft implies irregularity, picturesqueness, decorative value and uniqueness: figured textiles, pottery vases, decorative friezes, hand-wrought metal work, hand-hammered silver bowls. The machine implies precision, simplicity, smoothness, reproducibility: plain textiles, vases as simple as laboratory beakers, smooth polished metal work," Johnson wrote. "The craft spirit does not fit an age geared to machine technique."

There it was, an assertion slipped into one sentence that was practically a manifesto for a new way of thinking about machines: as tools that were not merely conduits for human agency, but tools whose use could carry distinctive characteristics that suggest something approaching a kind of agency of their own.

Barr and Johnson set out to lay a historical and theoretical foundation for the show—and, by extension, the International Style, and, by further extension, all of modernism. The press materials and catalog text began with a quote from Plato, an excerpt from *Philebus* printed first in ancient Greek and then in an English translation:

> By beauty of shapes I do not mean, as most people would suppose, the beauty of living figures or of pictures, but, to make my point clear, I mean straight lines and circles, and shapes, plane or solid, made from them by lathe, ruler and square. These are not, like other things, beautiful relatively, but always and absolutely.

In their coverage of the show, major news outlets reprinted this quote in its entirety: the *New York Times*, in three separate articles; the *New York Herald Tribune*, in two pieces; the *Los Angeles Times*; the *Christian Science Monitor* . . . Nearly every paper accepted the text on the museum's terms and spread the message that twentieth-century machines reflected values traceable to the court of classical aesthetics. Only a couple of writers pointed out that the conception of art familiar to citizens of the twentieth century was wholly unknown in Plato's time. (Plato considered geometry a discipline superior to art as he thought of it, and certainly never meant his appreciation for the beauty inherent in lines and circles to be a defense of their use in art as we think of it.)

"Machine Art" proved to be a publicity bonanza for the young Museum of Modern Art, garnering coverage in dozens of papers around the country. Writers and editors reveled in the high–low incongruity in the very idea of the exhibition, and poked gentle fun at it with such headlines as "Even the Kitchen Sink" (*Christian Science Monitor*), "Is Beauty the Same Thing in Auto's Lines and in Sunset?" (*Los Angeles Times*), and "Pots and Sinks Going on View as Art at Machinery Exhibit" (*New York Herald Tribune*). The majority of the reviews were laudatory, though, and seemed to reflect writers opening up to a new mode of thought about machines and modernity. "Already the machine has got rid of much that was not beautiful and is establishing new and true standards and values in articles of common use," wrote a writer (uncredited) for the *Chicago Tribune*. "Overdecoration and crude detail are giving way to simplicity, to better design, to a sense of finesse which is or should be an essential factor of taste. The New York exhibit was a remarkable demonstration of the beauty which is to be found in ball bearings, propellers, dictaphones, cash registers, weighing machines, kitchen utensils and such articles of common practical use, and alert makers are discovering that beauty has an important selling value."

The novelist and critic Robert M. Coates, writing in *The New Yorker*, pithily distilled the show's success at altering the terms of aesthetic evaluation. "It's disturbing, after all, to discover that you've been surrounded by beauty all your life and never known it," noted Coates.

Critics of the show, while few in number, tended to fault it sweepingly for an indifference to human feeling and disregard for intention. "The work of art presupposes the intervention of the artist in a very personal sense," argued Royal Cortissoz in the *New York Herald Tribune*. "In so far as he is an authentic artist, a truly creative agent, he leaves his thumbprint stamped ineffaceably, and delightfully, upon the work of art. Our pleasure in it springs not merely from his accurate handling of forms but from his emotion, his imagination, his whole character. That is what distinguishes the master from the tyro and, by the same token, that is what distinguishes art from 'machine art.'"

Before the end of 1934, Philip Johnson would resign his post as curator of architecture for the Museum of Modern Art "to pursue outside interests," in the museum's generously vague language. He walked away from the arts and took a hard right turn into political extremism, supporting the fascistic anti-Semite Father Charles Coughlin—an interest far outside the borders of the New York art world, as well as the boundaries of reason and principle. Contributing regularly to Coughlin's newspaper *Social Justice*, Johnson wrote unbridled anti-Jewish and pro-Nazi tracts, striking acts of intolerance and hate for a gay man with a Black lover. (Closeted at the time, Johnson would eventually come out and discuss the relationship he had with the actor and singer Jimmie Daniels in the mid-1930s.) Working on behalf of Coughlin's organization, the National Union for Social Justice, Johnson helped form a group for young Coughlin supporters. He no doubt took inspiration from a Nazi youth rally he had attended in Potsdam. As he would later admit to one of his biographers, Franz Schulze, "You simply could not fail to be caught up in the excitement of it, by the marching songs, by the crescendo and climax of the whole thing, as Hitler came on at last to harangue the crowd." Johnson was more than sympathetic to Hitler, at one point writing an admiring review of *Mein Kampf* for the *Examiner*, a Connecticut quarterly.

Machine art was nonhuman. Fascism was inhuman. The two are not the same. And yet, one cannot help but wonder if a select few of the qualities associated with machines in the modern era—cold efficiency, severity, suitability to mass replication—spoke with special resonance to a fascist like Philip Johnson.

5

SPIRITUALS OF THE CITY

The workings of the industrial world shifted from mechanics (gears, springs, and pinions) to electronics (currents, tubes, and radio waves) early in the twentieth century, and arts of many kinds were transformed, with accelerating progress in forms that benefited most from technology, such as sound recording, motion pictures, and broadcasting. The phenomena of recordings and movies actually predated the electric revolution, with wind-up Victrolas for playing recordings made with sound cones rather than microphones, and silent films made with hand-wound cameras and the sun as the light source rather than incandescent bulbs. Still, electricity quickly proved to be more than a new source of power to turn wheels and drive pistons.

Among other things, electricity made possible a machine of a new sort for making music: the microphone. It was a machine for the electrical age, whose moving parts were electrons, flying on the microscopic level. As with all electronic machinery, its central workings were internal, a parallel to the mind—and the soul—by extension of our conception of mechanical machines as parallels to the body.

This new type of machine did things of a new kind, augmenting and altering the work of humans in ways that changed the art they were practicing. It transformed popular music, helping to make it one of the dominant artforms in the electric age. "One thing that was tremendously important was learning the use of the microphone," Frank Sinatra said, describing his tutelage as a vocalist in the early decades of radio and recording. "Many singers never learned to use one. They never understood, and still don't, that a microphone is their instrument. It's like they are part of an orchestra, but instead of playing a saxophone, they're playing a microphone."

When commercial radio broadcasts and electric recordings began, in the second decade of the twentieth century, popular music was an art defined by its long-standing system of distribution: by performers on stage, belting out songs to reach the back rows of the balcony without the benefit of microphones or amplification. It would take a new generation of singers to grasp and adapt to the potential for communicative intimacy made possible by the microphone. Singing softly, naturally, with the tonalities and inflections of ordinary speech, singers could speak directly in a one-on-one way to listeners alone in the privacy of their homes, not surrounded by audiences in public theaters or concert halls. Before long, songwriters would pick up on the possibilities of the new style of "crooning" and write words and music suitable to the intimacy that the microphone and the home environment allowed: songs of secret longing, confidences, and solitary reflection.

The first important innovators of this new aesthetic were a group of now-forgotten women: in alphabetical order, Betty Brown, Glory Clarke, Gertrude Dwyer, Nancy Foster, Gloria Geer, Sadie Green, Mamie Lee, Angelina Marco, and Marion Ross. They shared the same sensitivity to musical nuance and subtlety of technique, the same deft mastery of the tools of electronic transmission. They even shared the same vocal range and timbre, because they were all the same person: a versatile and far-sighted vocalist whose real name was Vaughn de Leath and who recorded under that name, too, to a degree of acclaim at one time. A former music student at Mills College, de Leath was a sophisticated musician gifted with a wide range, from deep contralto to bright soprano. At some point in 1919,

Lee de Forest—a New York-based inventor and entrepreneur—heard about de Leath and invited her to perform on one of the first broadcast facilities in the U.S.: his experimental radio station, channel 2XG (later licensed by the Federal Communications Commission as WJX), housed in a small space in the World's Tower, a high-rise a few blocks south of Times Square on Broadway. This was in January 1920, and the carbon microphones that constituted the most advanced audio technology of the day were built with delicate, handmade vacuum tubes that could blow out if subjected to sound that was too loud or too high in the frequency range. De Forest's station was the first in the country to use them, because de Forest happened to be the inventor of the vacuum tube integral to early radio broadcasting and signal processing more broadly—no small achievement.

De Leath warmed up, preparing to sing in the impressive upper register of her voice. When de Forest warned her of the risk to his equipment, she modulated down to contralto and proceeded to sing one of the hits of the day, George Gershwin's "Swanee," at conversational volume, as de Leath would recall the event. "She was an instant success," wrote de Forest in his autobiography, describing the impact of a performance a fairly small number of people could have heard. "Her voice and her cordial, unassuming microphone presence were ideally suited to the novel task. Without instruction she seemed to sense exactly what was necessary in song and patter to successfully put herself across." It was significant that this young white woman proved deft at delivering a song derived from Black expression such as "Swanee," which Al Jolson had made famous, singing in blackface, and which player pianos had further popularized as a piano roll. De Leath would record very little of such material and would never be closely associated with it, but she could sing it with tempered feeling suitable to the airwaves.

If not precisely "instant," de Leath's success was substantial in the primordial days of radio. "A flood of fan mail . . . testified as to the important broadcasting success which she had here suddenly achieved," de Forest wrote. "Thereafter she appeared frequently before that early microphone." In the distinctive performances that made her reputation, she sang in a confident but restrained, unaffected, untheatrical voice, in the cadences of everyday speech. Performing on an early radio station in Newark, New Jersey, that would become New York's WABC, she was lauded as "the Radio Girl" and "the First Lady of Radio." Recording for no fewer than eight record labels, de Leath made hits of songs that exploited electronic recording and delivery to connect directly with someone listening alone, including a ballad Elvis Presley would murmur thirty years later, "Are You Lonesome Tonight," along with a winking nod to transgressive sexuality, "He's So Unusual." (Al Jolson could never have belted "Are You Lonesome Tonight" to a theater audience of a thousand people, none of whom would have been lonesome that night.)

A composer as well as a singer, de Leath was said to have written hundreds of songs, including a collaboration with Irving Berlin ("Drowsy Head"), done when he was still primarily a lyricist. She played multiple instruments—piano, guitar, ukulele, and banjo—and was sufficiently skilled at both music and business to host her own radio show, lead its sixty-piece orchestra, and manage the station, a short-lived AM operation in New York. The circumstances of her rapid decline and death at forty-two are unclear; she was known to drink, and her stocky body type would have limited the roles available to her in films, if she even wanted to act. In 1930, a wire service story looked back on de Leath's career with the headline, "'Original Radio Girl' Won Fame with Crooning Voice." Vaughn de Leath was all but forgotten already, though the approach to the microphone she innovated had become the standard for popular singers in the era of electronic delivery. Along with de Leath came Jack Smith, "the Whispering Baritone"; Chester Gaylord, "the Whispering Serenader"; Art Gillam, "the Whispering Pianist"; Nick Lucas, "the Crooning Troubadour"; and dozens more, from Annette Hanshaw and Gene Austin to Rudy Vallee on to Bing Crosby and then Sinatra and Billie Holiday (see photo, page 62) and countless others until rock 'n' roll came along.

An expanding set of electrical machines—microphones, radios, record players, and the vast array of amplifiers, transmitters, record cutters, and miscellaneous arcana to facilitate and interconnect everything—made a whole new approach to a popular art possible. Through new technology, the singing of songs grew quieter, more intimate, more personal. Machines helped make an art more human, while critics of the new technologies took them as dehumanizing. As mechanisms for creating music in new ways, the microphone and amplifying equipment were derided by critics for promoting *wrong* ways. Instead of using the power of the lungs and physical skills to project full, round tones from the diaphragm, crooners crooned. Electronics, in making new creative options possible and altering the terms of an art, were thought of as tricks for cheating.

Protectors of the public good bridled at the sound of male crooning as a threat to the masculine order. "I desire to speak earnestly about a degenerate form of singing which is called crooning," announced William Cardinal O'Connor of Boston in an address to Catholic men in the Holy Name Society in 1932. "No true American man would practice this base art. Of course, they aren't men. I can't turn the dial without hearing those whiners, crying vapid words to impossible tunes. If you will have music, have good music, not this immoral and imbecile slush."

Executives of radio companies, more accommodating on a matter good for their business, provided songwriting guidelines to composers, urging them to employ no more than five notes on either side of middle C, to suit broadcast equipment. Both male and female singers pared down and turned inward, altering the public image of men in popular music. "The male star prototype softened, melting from He-Man into Dream-Boy," in the words of the writer and musician Ian Whitcomb. Technological change can trigger social change, sometimes for the good.

Once electrified, the mechanisms for recording and reproducing music grew progressively sophisticated and elaborate, making possible ways of manipulating sound that altered how music could be made, heard, and understood. John Cage advocated this presciently, calling as early as 1937 for a new "twentieth-century means for making music" employing "the new materials, oscillators, generators, means for amplifying small sounds, film phonographs, etc." In both popular music and the realm of Cagean experimentation, electrical machinery changed what music *was*, or what it was conceived to be; and the advent of digital technology in the second half of the twentieth century changed it once more. An ever-growing assortment of recording technologies gave makers of recorded music the ability to incorporate ambient sounds or mechanical sounds— any noises or "nonmusical" sounds—in the process redefining them as musical. Editing, overdubbing, signal processing, sound synthesis, and other techniques expanded the textural range and the expressive capacity of recorded music. Seagulls squawked in the sonic landscape as the

Shangri-Las went walking in the sand. Chickens clucking "good morning" for the Beatles morphed into an electric guitar.

The increasing aural sophistication of recorded music called for commensurately advanced home playback gear, just as the equipment—a status symbol for the postwar generation of young people—called for increasingly sophisticated music. *Sgt. Pepper's Lonely Hearts Club Band* begat shopping centers full of stereo systems, and rec rooms full of stereo systems begat collections of hi-fi records. *Sgt. Pepper*, a totem to the recording studio as an instrument, if not a full electronic collaborator—the sixth Beatle, after producer George Martin—provided benefits nearly as varied as the cutout faces on the album cover. It gave the Rolling Stones something to riff off with *Their Satanic Majesties Request*; it gave acolytes like Pink Floyd and King Crimson something to build from and take to extremes with grand monuments of pop pretentiousness; it gave Black musicians like Charles Mingus and Quincy Jones something to reject as white bullshit; and it gave acoustic singer-songwriters something to play against. "Stephen [Stills] and Graham [Nash] and I heard all the magic tricks the Beatles were pulling in the studio, and that was a big inspiration to us," recalled David Crosby. "We decided to do the exact opposite. Fuck that shit." The gadgetry of the EMI studios on Abbey Road, where the Beatles recorded, accomplished more than it was designed to do.

By the late 1960s, an inversion of the function of recording was complete, though not all-encompassing, in popular music and the avant-garde. Records were not solely documents of performances by musicians playing instruments—records in the traditional sense of record-keeping. The recording had become the artform; it was now the music, not something *done to it*. The recording studio, instrumental in the making of that music, was now tantamount to a musical instrument. Much as Frank Sinatra had taught singers of an earlier era that the microphone was their instrument, music makers of the late twentieth century learned to treat the studio as theirs. Live performers, in turn, tended to try to reproduce the sound of their recordings on stage, augmenting live instrumentation with electrified effects and, increasingly, prerecorded elements—including, sometimes, "guide" vocal tracks that the singers would sing over (or, on occasion, lip-sync to). In many cases, the sound systems of venues would be calibrated to emulate the processed audio quality of recordings. The aesthetic of studio technology was infusing the music, in and out of the studio.

In the early 1980s, one of the key figures in the American hi-fi industry, Avery Fisher, gave a talk at the Consumer Electronics Show, which I was covering as a tech journalist in one of my first jobs out of college. He led off with a story about having recently attended the concert venue at Lincoln Center named for him, Avery Fisher Hall, after he had donated a sizable portion of his fortune from merchandising Fisher-brand stereo components to fund renovations in the hall to improve its acoustics. He had gone to see a rock band. I can't remember the name of the group, but I have never forgotten how Fisher described his experience to an audience of electronics industry executives. "My God," he said, "they're trying to sound like *us*." The rock act, performing live, wanted to sound like a record playing on a stereo.

WITH THE GREAT MIGRATION IN THE YEARS AFTER THE FIRST WORLD WAR, more than two million Black Americans left the South in hopes of finding better work and making a better life in the urban, industrial North, and they brought their music with them. Muddy Waters, born McKinley Morgenfield, grew up in Mississippi and moved in young adulthood to Chicago, and so did the blues. The same was the case for James Cotton, Willie Dixon, Honeyboy Edwards, Jimmy Reed, Howlin' Wolf, and other musicians whose music would come together by midcentury as a new school of work: electric blues. "I followed the money, went where there was jobs," recalled Edwards in an interview I was fortunate to do with him in a boardinghouse in Chicago in 2003, the centennial of the date ascribed to W. C. Handy's "discovery" of the blues. Edwards was eighty-eight and would continue performing until his death seven years later. "All there was was field work in the Delta, hard labor, and I didn't want nothing to do with that," he said. "I went around with Robert Johnson, this town and that, and after he died that day, I said, 'I'm getting out of this place now. There's nothing for me here no more.'

"I came up north, and everything was loud and noisy. I kept on

playing, but I had to change it around now. That's what we did. We made the music loud and noisy, just like this place [Chicago]."

An art of anguish and joyous release, and of shared experience relayed in multilayered, often coded language, the blues first took form in community spaces and private quarters—in storefronts, on porches, in parlors, and outdoors in Black communities throughout the South during the first decades of the twentieth century. It was sung and played primarily at first by solo musicians and small groups using acoustic instruments—affordable, portable, and replaceable instruments, typically guitars. By midcentury, guitar manufacturers had begun making electrified models with solid wood bodies (or thin semi-hollow bodies), which could be factory-produced at a lower cost than hand-luthiered guitars and marketed as starter instruments. Classical and jazz musicians tended to look down on these guitars, deriding them as toys for amateurs. (Most notable among the artists early to defy this bias were Charlie Christian, who played brilliantly on the electric guitar for Benny Goodman, and Christian's mentor, the jazz arranger and multi-instrumentalist Eddie Durham.) For the wave of Delta blues musicians arriving North around the time of the Second World War and soon after, the new electrified instruments solved a practical problem, while fitting in neatly with the surroundings. "When I went into the clubs, the first thing I wanted was an amplifier," Muddy Waters would remember. "Couldn't nobody hear you with an acoustic."

As the critic Francis Davis described this scene, "South Side clubs . . . must have been as clamorous as the factories in which black clubgoers earned the dollars they parted with." Following the money, as well as the social capital and peer companionship in their new setting, blues musicians found work in the bars and clubs, at rent parties, and in after-hours spots throughout Chicago's Black neighborhoods—loud and noisy places to commune and seek relief from the daytime grind and white bullshit in a loud and noisy city with limited opportunities at unlimited cost for people of color.

The author Richard Wright captured the numbing, disorienting shock of urban migration in his book *12 Million Black Voices*. "It seems as though we are now living inside of a machine; days and events move with a hard reasoning of their own," he wrote.

> We live amid swarms of people, yet there is a vast distance between people, a distance that words cannot bridge. No longer do our lives depend upon the soil, the sun, the rain, or the wind; we live by the grace of jobs and the brutal logic of jobs. We do not know this world, or what makes it move. In the South life was different; men spoke to you, cursed you, yelled at you, or killed you. The world moved by signs we knew. But here in the North cold forces hit you and push you. It is a world of *things* . . . [Italics and ellipses in original text.]

The blues that musicians carried north and reinvented carried Black vernacular music's traditional power of succor, translated electrically and amplified for use in Chicago. "The blues could be called the spirituals of the city," Wright noted. It was (and to some degree remains) a music for "people whose life has been caught up in and brutalized by the inflexible logic of modern industrial existence."

Muddy Waters's first hit record, "I Can't Be Satisfied," released in 1948, was a resetting of a composition he had recorded seven years earlier in Mississippi for the Library of Congress under the direction of the folklorists Alan Lomax and John Wesley Work III. In its first life, it had been a spare but propulsive, musically intricate piece for voice and acoustic guitar titled "I Be's Troubled." Playing slide guitar, gliding across the frets on the neck with a knife blade or sawn-off neck of a bottle, Waters produced a chug-chugging rhythm while interjecting counterpoint lines picked on individual strings, all while singing a catchy song about dissatisfaction.

In the Chicago version, Waters used an electric guitar built with a slab of magnetic metal for a pickup. It had not been designed for bottleneck playing. The crude pickup and closely miked studio amplifier exaggerated and distorted the slide effects, giving the record a wild, novel sound. Now widely recognized as the first hit recording with electric slide guitar, "I Can't Be Satisfied" captured Black dissatisfaction in the Jim Crow North with gnarly electrical bite on a recording neatly complicated with a driving, danceable rhythm and a vocal delivery laced with an incongruous tinge of pleasure, a hint of the possibility of satisfaction, somehow, after all. The record was an original statement of a different kind of blues, an electrified spiritual for an electrifying city. Its quality and importance should not be measured by the fact that a group of young, white Englishmen would one day become Muddy Waters fans, start a band named for one of his songs, "Rolling Stone," and get the idea to write a song of their own about not getting satisfaction.

Electrified and amplified, its sound distorted by musicians and singers experimenting and pushing things past the red safety line, the blues that took form in the urban North was aggressive and dynamic: lacerating. Though drawn from the more laconic, countrified acoustic blues of the South, the new music had a kick and swagger that seemed to capture something changing in Black America. The insistent beats and amped-up volume of the music also made it well suited to dancing—a lacerating music that was a unifying social force. "We excited the people," Honeyboy Edwards said. "We shook up their heads and got them moving."

Dozens of Chicago musicians recorded high-voltage blues in the late 1940s and early 1950s, sometimes sharing studio musicians or trading songs to build a collective body of work with insistent rhythms and anxious electric energy to reflect the Black experience in Northern cites of the postwar era. "A lot of things changed when we got up North," recalled James Cotton, a native of Mississippi whom Muddy Waters called up to Chicago to join him as his bandleader and harmonica player in 1955. "Look here at the harp," Cotton said, pointing to the blues harmonica that was an integral part of the Chicago sound. "Mississippi—no mike. The harp was the sound of a train coming into town, that big machine in the distance somewhere. You blow soft and quiet now, and that train goes off down the tracks. Now, here, [at] Chess [the record company]—you have the harp right on the mike—you're holding them together, tight— and that changes things. You're blowing right into the electricity. Now it sounds like you're there inside the machine. You *are* the damn machine."

The harmonica has a set of qualities that have long made it good for imitating trains: a timbre close to the tone of an engineer's whistle, the ability to make chugga-chugga rhythms by inhaling and exhaling rapidly over the reeds, and a facility for playing in glissando by cupping the instrument tightly and slowly opening or closing the palm of one hand, a technique that simulates the Doppler effect of pitch rising or falling as a train comes closer or rides away. Back-of-mind associations can come into play, too, and connect the sound of the harmonica to the visual cliché of a hobo riding the rails, sitting in a boxcar playing a mournful blues on his harmonica.

More tellingly, though, the sound of locomotives connected to mid-century popular music in a way that points to the pervasive influence of various modes of mechanical transportation on the music of their time. From the rise of the Machine Age to the current day, the music of every period in history sounds something like the transportation machinery of its age. This is more of a theory than a hard-and-fast rule, but it's an illuminating one. In the swing era, the steady 4/4 beat of train wheels clicking on railroad tracks maps elegantly over the drumbeat on core records of the time, such as "Stomping at the Savoy," "One O'Clock Jump," "In the Mood," and "Take the 'A' Train.'" (That last number, composed by Billy

Strayhorn for the Duke Ellington Orchestra, was about a New York City subway line that went to Harlem, not an aboveground locomotive, though the beat so suits aboveground trains that the motion picture Ellington made of the song in the 1940s is set in a standard railcar and begins and ends with the sight and sound of a barreling locomotive.) The ostinato rhythm section arrangements on these recordings echo the clickety-clack, clickety-clack of train tracks, and it's not too much of a stretch to think of the riffs of the horn sections as accents akin to train whistles.

In the days of early rock, when automobiles replaced railroads in the popular imagination as well as the American landscape—and, for young people in particular, took on special meaning as a symbol of generational independence and freedom—countless songs dealt explicitly with cars and driving with words that blurred past like road signs on the highway: "Riding along in my automobile, my baby beside me at the wheel . . ." The record generally acknowledged as the first rock 'n' roll song (a designation endlessly if pointlessly contended) was named for the make of a new model Oldsmobile, "Rocket 88," and was all about the power of a big, fast eight-cylinder car—mechanical but sexual, mass-produced but the essence of individual power to the lone driver on the road. Musically, moreover, the songs sounded like the cars they were written about, with backbeats like the rumble of eight cylinders and electric guitar chords growling from amps like the engine of a Pontiac revving up. "Ambition is a dream with a V8 engine," said Elvis Presley or one of his press agents.

By the 1960s, America and its music had entered the jet age, only to leave that early for the space age. America's answer to the Beatles, the Byrds, had originally been called the Jet Set, and their leader, Roger McGuinn, explicitly conceived of the rich, soaring sound they achieved through dense vocal harmony and McGuinn's twelve-string electric guitar as a parallel to the sound of jet aircraft. As he was quoted in the liner notes to the group's debut album, *Mr. Tambourine Man*, "The sound of the airplane in the forties was a rrrrrrrrooooooaaaaaaaahhhhhhhhhh sound and Sinatra and other people sang like that with those sort of overtones. Now we've got the krrrriiiiisssssssshhhhhhhhhhhhhh jet sound, and the kids are singing up in there now. It's the mechanical sounds of the era."

In little more than a year, McGuinn and his bandmates would shoot higher in every sense, going psychedelic with "Mr. Spaceman" and "Eight Miles High"—songs that used the imagery of space travel and sonic effects to conjure a feeling of druggy travel through the inner spheres. "We got trippy, man—the ultimate trip through our minds," recalled David Crosby, McGuinn's bandmate in this early iteration of the Byrds. "I don't think any of us [in the band] had actually tried LSD yet or really done any drugs to speak of. We were full of shit. But we were good at playing the part, and we caught up to the music pretty quickly. My head caught up with my guitar."

Just as the sounds of trains, cars, planes, and rocket ships seeped into popular music, ambient sounds from industrialized America—mechanical ambience—influenced some important musicians of the second half of the twentieth century. Where one person would hear noise and find it horrifying, another could hear noise, as well, and find it thrilling. A case in point took place in a Ford plant in Michigan in 1955, when rock 'n' roll was beginning to break through the airwaves on a national level. Carpenter Elementary School, which served the small town of Ypsilanti as well as part of neighboring Ann Arbor, organized annual field trips to the nearby River Rouge complex in Dearborn, once the largest industrial facility in the United States, where Ford would one day manufacture Mustangs. (See photo, page 70.) One of the young students visiting the complex was a third-grader named Jim Osterberg, who was growing up in the Coachville Gardens Trailer Park on the outskirts of Ypsilanti.

"I remember being led up on one of these high catwalks and watching large pieces of sheet metal basically dropped in a very organized way from a high height through a chute in such a way that when they hit bottom they would hit a mold, which would form them into something that would begin to resemble a door. But it was the sound it made. That had a big effect on me," he would recall decades later, after changing his name to Iggy Pop.

"*Whe*wwwww!" he said, imitating the sound of the sheet metal crashing onto the mold in the "giant metallic castles, these castles of hell" at River Rouge. "I wanted to make music, and I thought it should sound like that. And I loved it. It was so impressive. It was *power.*"

As much as any electric blues musician was taken in by the

high-wattage atmosphere in Chicago, Iggy Pop found himself caught up in the industrial sensibility of the Detroit area. The city with a motor in its nickname had the sound of heavy machinery in the air. "Detroit has a beat—the pounding out of fenders, the pounding of bumpers, the day-by-day grind that made us," said the veteran Detroit broadcaster and educator Russ Gibb. "You had to have the beat because even on the line, things came through with a rhythm. Every three or four minutes, that line would move, and you'd have to pound on the hubcaps. There was always a rhythm to Detroit."

The producer and music mogul Berry Gordy, a onetime production worker at a Ford plant, grew famous for taking lessons from the assembly line and applying them to Motown, the vertically integrated music business he ran. "Those slow-moving car frames were the loveliest sight I'd ever seen," he wrote in his memoirs. "There was a pleasing simplicity to how everyone did the same thing over and over again."

When he started Motown, he would recall, "My own dream for a hit factory was quickly taking form, a concept that had been shaped by principles I had learned on the Lincoln-Mercury assembly line. At the plant, the cars started out as just a frame, pulled along on conveyor belts until they emerged at the end of the line. I wanted the same concept for my company, only with artists and songs and records. I wanted a place where a kid off the street could walk in one door an unknown and come out another a recording artist—a star."

The "power" Iggy Pop heard and felt at River Rouge in third grade was unrelated to the economic independence and force of creative and social aspiration that Berry Gordy sought. Pop's fascination was with destructive, not constructive power, though his attraction to loudness and noise was not a perverse interest in ugliness. No—he found beauty in noise. He loved it, and in the music he made and through the music he influenced others to make, he taught countless people how to hear, think, and feel more expansively, to find music in the noise.

He reveled in the mechanical sounds of everyday life, taking noises he heard in childhood as extra-musical epiphanies: the buzz of his father's electric razor; the ratchety hum of a small, cheap space heater in his family's trailer, a sound he would always treasure as "my first musical gift from God"; the swoosh of cars zooming down the highway by the trailer park. After a conventional garage-rock apprenticeship, playing drums in various cover bands—including one, the Iguanas, that gave him his nickname—he put together a four-person group that included two brothers, guitarist Ron Asheton and drummer Scott Asheton, whom Pop chose specifically because he liked that they could barely play their instruments.

Gigging in small venues in and around Detroit, the band justified its original name, the Psychedelic Stooges, by making freeform experimental music with a vibe of cheeky, arty shock value. They used household appliances along with their instruments to create an aural atmosphere of droning chaos, with the mics and amps set up to produce more feedback than sound. "Before we were making records, we were making a big avant-garde mess," Iggy would recall. "We would show up with some oil cans and vacuum cleaners and egg beaters, and also electric rock instruments, and we would play a kind of trance music."

As the Stooges, Iggy Pop and his group brought the ravaging spirit of River Rouge's castles of hell to the rock clubs and rec-room stereo systems of America. Though the band's early records never sold in Top 40 numbers, the Stooges and their notorious stage shows—the musicians thrashing as their leader, bony and bare-chested, exposed himself and tore gashes into his skin with broken glass—brought rock into the domain of performance art. Critics fawned over the Stooges' "driving monotony and terrifying intensity," and a generation of musicians and listeners would learn to love the noise in music that would become codified as punk, metal, noise rock, and industrial music.

"Not much in that music speaks to me as a musician," said John Cale, the composer and cofounder of the Velvet Underground, who produced the Stooges' first album. "But I can appreciate it as a man, without much personal experience in the area of America Iggy came from. As a Welshman, I know the life of the miners and the factory workers in South Wales. I can hear the sound of the steelworks of Shotton in the music of the Stooges I recorded.

"It was hardly music at all. It was mainly noise. But it was powerful. It was frightening."

6

THIS IS MUSIC?

The composer weighed five tons.

Called the Illiac, short for the Illinois Automatic Computer, the first machine to use electronic computing technology to create music was ten feet tall and filled nearly a whole room in the science facility of the University of Illinois Urbana–Champaign. To build it at a time when no other educational institution in America had its own computer, the trustees of the university had to petition J. Robert Oppenheimer, director of the Institute for Advanced Study in Princeton, for permission to use a high-level internal architecture developed at the IAS by the Manhattan Project mathematician John von Neumann. Revered for doing brilliant calculations in nuclear physics, von Neumann was also notorious at Princeton for blasting records of rousing march music on his phonograph while he worked, annoying another scientist with a desk near his, Albert Einstein. It is a mere coincidence, but still a fitting one, that the Illiac, the first computer to learn how to compose, had a design attributed to a mathematician who marched to music in his head.

Completed and put into operation at the University of Illinois in September 1952, the Illiac was able to read data provided in the form of punched tape, store it in its memory, and perform up to 11,000 arithmetic operations per second. It could function for up to eight hours per day but no longer, in order to preserve the working life of its 2,800 vacuum tubes. The Illiac was in heavy demand by faculty and students doing work in physics, astronomy, and other disciplines during the Cold War boom in science research. For instance, a chemistry professor, Dr. Lejaren A. Hiller Jr., was working under a government-sponsored grant from Dupont to study the physics of synthetic rubber, using the Illiac to calculate the macromolecular dimensions of cellulose. "I had an idea one day when I was hanging around the chemistry lab just doing I don't know what, when I thought, 'Well, you know, if I change the geometrical design of this random light program I've written,' which had gotten quite complicated, 'change the parameters—the boundary conditions, so to speak—I can make the boundary conditions strict counterpoint instead of tetrahedral carbon bonds.' And that's how it all started."

Hiller, like von Neumann, was a scientist with a taste for music. Born and raised in Manhattan, he grew up exposed to science and art in creative juxtaposition, through the unusual work of his father, Lejaren à Hiller. (That's a lower-case "a" with a reverse accent mark, which the elder Hiller added as a young man when he moved from Wisconsin to New York to study art and changed his name from John to Lejaren.) A photographer and painter, Lejaren à Hiller became renowned in the commercial art world for staging elaborate tableaux with sets and models in costume for use in photographic illustrations, such as a series of reconstructions of surgical procedures through history which he created for a medical supply company. His wife, the former Sarah Plummer, had been in the chorus line of the Ziegfeld Follies (a Ziegfeld Girl) and had modeled for Charles Dana Gibson, whose pen-and-ink drawings of demure, highly polished women were taken as an ideal of white womanhood (the Gibson Girl) in the early twentieth century. Their son would remember his parents throwing "wild parties with nude women and models running around the house," sometimes leading to raids by the police.

The younger Hiller took piano lessons as a child and created his first original music for the family's player piano, reworking the piano rolls by "cutting designs and punching holes" in the paper, adapting the mechanical music technology to create sounds he thought of as "highly satisfying effects." This, with no knowledge of Conlon Nancarrow. At college, in Princeton, he focused on science, getting his bachelor's and first master's degree, as well as his PhD, in chemistry, while also studying music with Princeton's marquee iconoclasts, Milton Babbitt and Roger Sessions. By 1947, he had composed a piano sonata and written his dissertation, "The Chemical Structure of Cellulose and Starch."

As a junior professor at the University of Illinois, Hiller carried on his chemistry research as he acted on the lightbulb realization that he could get the Illiac to compose music by simply changing the geometrical design of the random light program. A photo from this period in the university archive (see page 78) captures Hiller as the very image of Eisenhower-era industriousness, seated in a dark suit and wearing black horn-rims, surrounded by Fifties technology—an oscilloscope, half-inch tape decks, an array of patch plugs—as he counts time on a stopwatch with his left hand and turns a knob to adjust a VU meter with his right. Colleagues would remember Hiller for a puckish sense of humor that escaped the photo record.

He took up the project of coaxing the Illiac to compose music as "a bootleg job at night," he said, and enlisted a graduate assistant in the chemistry department, Leonard Isaacson, to help with the programming. The two of them put about two hundred bootleg hours into the process of inputting instructions on punch tape for the Illiac to work from. They adapted a technique that had already surfaced in this early stage in computer processing: the Monte Carlo method, which exploited the ability of a computer to take in a great number of random terms and test them against a set of conditions or rules. As Hiller would explain in a book he cowrote with Isaacson, *Experimental Music: Composition with an Electronic Computer*:

> Music is a sensible form. It is governed by laws of organization which permit fairly exact codification. From this proposition, it follows that computer-produced music which is "meaningful" is conceivable to the extent to which the laws of musical organization are codifiable.

If it was good for Bach, it would be good for the Illiac. Computer music that would make sense to most listeners would have to follow the same rules those listeners were accustomed to hearing in music made by human beings.

Further, Hiller and Isaacson explained:

> It is a feature of digital computers that they can be efficiently used to "create a random universe" and to select ordered sets of information from this random universe in accordance with imposed rules, musical or otherwise. Since the process of creative composition can be similarly viewed as an imposition of order upon an infinite variety of possibilities, an analogy between the two processes seems to hold, and the opportunity is afforded for a fairly close approximation of the composing process utilizing a high-speed electronic digital computer.

That is to say, the human composer's process of applying rules to a set of options could be transferred to a computer capable of handling a great many options. The processes are similar. But, one might ask, don't human composers do more than follow rules? Hiller and Isaacson recognized this and did some things to address matters of judgment and taste, though they stopped short of dealing with an even thornier consideration: inspiration. As they wrote:

> It should be noted . . . that the composer is traditionally thought of as guided in his choices not only by certain technical rules but also by his "aural sensibility," while the computer would be dependent entirely upon a rationalization and codification of this "aural sensibility."

Their limited method of approximating certain qualities of sensibility was to introduce a set—or a set of sets—of standards for the Illiac to apply. Hiller and Isaacson gave the suite a four-movement structure, with the sections intended to grow progressively complex musically. The first movement would begin by using the linear counterpoint rules of Renaissance music to offer up simple melodies, and then bring in more musical perspectives or "voices"; the second movement would fully employ the "four-voice" approach to composition that first-year conservatory students learn—the process of composing musical lines that work to complement or contrast one another; the third movement would introduce denser chromatic harmony, updating the sound and complicating things considerably; and the final, fourth movement would be "a complete departure from traditional compositional practice"—anything goes, or at least anything the Illiac could do using the Markov chain process of random (or pseudo-random) selection.

At the end of the process for composing each movement, the computer produced a sequence of numerical expressions that Hiller converted to conventional musical notation on standard music paper, working by hand. Translation of those written scores into audible sounds then required the use of living humans playing musical instruments: a string quartet.

With the fourth section not yet finished, a three-movement version of *The Illiac Suite* had its premiere on August 9, 1956, as part of an Illini Composers series of concerts at the University of Illinois. A full house of some two hundred people attended the performance by a quartet of undergraduate musicians in the Wedgewood Lounge of the student union building. The event "had a very electric atmosphere," Hiller would recall, "because there were a lot of people who were very resentful on general grounds." From the moment the performance began, Hiller said, "some people had already become livid with rage over it."

Within days, panicky coverage of the event began appearing in newspapers around the country. Viewpoints toggled between skepticism and cynicism in pieces typified by a *Chicago Tribune* story titled "This Is Music?" The writer, one William D. Hansen, warned readers, "Little by little, it seems the electronic 'brains' are taking over. They can solve complex mathematical problems father than humans can, they can translate books from one language to another, they can even tell subordinate machines what to do and chide them when they don't do it right. As if this weren't enough, two chemists at the University of Illinois now have taught a precocious electronic digital computer named Illiac to compose music. . . . While human composers are not yet in jeopardy, the cards are on the table."

In the *New York Times*, social commentator Brock Brower was more decisive. Under a blunt headline, "These Thinking Machines Can't Really Think," Brower pronounced, "To stretch the point as far as some of the computer people have done, machines are presumably capable of 'creating works of art.' . . . This rather ludicrous extension of the machine–brain equation to artistic creativity perhaps best illustrates its limitations. No machine is ever really likely to contain the artist within its electrophysics."

For Lejaren Hiller, the response to *The Illiac Suite*, particularly among the many people who had not heard it, was unsettling. "It was really a very strange summer, because I went from total obscurity as a composer to really being on the front page of newspapers all over the country," he recalled. "One week I was nobody, and the next week I was notorious." The authenticity police had raided the party.

"I can remember one person from a humanities department coming up to me on the street and saying that I was doing a terrible thing, that I was the worst person to turn up since Cesare Borgia, that I was an anti-humanist and I was destroying human uniqueness. I was burned in effigy. All of which is fun for a composer, really."

～～～

THE MAKING OF *THE ILLIAC SUITE* WAS A SIGNAL MOMENT IN THE ENLACED histories of art and technology, the point when it became indisputably clear that computers could have a role in the creation of at least one art, classical music. Less clear was the degree of agency attributable to the machine. Hiller, a lettered empiricist and creative adventurer in one horn-rimmed package, saw the matter in multiple ways. At times, he talked of his method of working with the Illiac in cooperative, some-what collaborative terms—as "conversing with the computer," much as musicians often describe how their instruments influence or inspire their music-making. At other times, he acknowledged the significant limits of the computer technology of the day, while also suggesting that all work we see as the product of creative inspiration could, hypothetically, be quantified as computer code. "Being 'creative' would seem to depend at the very minimum, like 'thinking,' on having a computer operate on a self-sustaining basis, and to 'learn from experience,'" Hiller wrote. "The pursuit of knowledge which depends on 'creative thought' is a technique of coding, of finding explicit statements of more and more complex logical relationship."

In the 1950s, few people in the general population knew much about computers, beyond the fact that these gargantuan things with peculiar, science-y names were suddenly in the news. Military needs had facilitated advances in computational technology during the Second World War, when Alan Turing applied his foundational ideas about computation to build the Colossus machines (kept secret until the 1970s) to help the Allies decipher Nazi codes, while the Third Reich had the Z3 computer built to calculate the trajectories of missiles intended to take out Allied equipment like the Colossus. In ostensible peacetime, Cold War science produced the ENIAC (Electronic Numerical Integrator and Computer), whose sole original functions were military, and its successor, the Ordvac (Ordnance Discrete Variable Automatic Computer), which was built for the Ballistic Research Laboratory at Aberdeen Proving Ground in Maryland. In fact, the Illiac at the University of Illinois was a precise duplicate of the Ordvac, able to run the same programs and communicate with it over a proprietary telephone line. By the late 1950s, the construction of mammoth, room-size computers was looking like a military-industrial craze. Beyond the ENIAC, the Ordvac, and the Illiac were the Univac (Universal Automatic Computer), the CSIRAC (Commonwealth Scientific and Industrial Research Automatic Computer), the MANIAC (Mathematical Analyzer Numerical Integrator and Automatic Computer), the Silliac (Sydney Illiac), and the Johnniac (John von Neumann Numerical Integrator and Automatic Computer). When DC Comics introduced a new android villain with a computer brain to battle Superman, in 1958, what else could he be called but Brainiac?

The making of music and art was generally understood to be a tangential application of these computers, though at least one creative thinker had seen the potential for art-making in computational devices as early as the mid-nineteenth century, when the English inventor Charles Babbage conceived of a proto-computer he called the Analytical Engine, and the mathematician and theorist Ada Byron, Countess of Lovelace, recognized that an engine for analysis could, hypothetically, serve as machinery for creation. In exhaustive notes she wrote on the Engine, first published in

1843, Byron presented the first ever case for making music with a computational machine:

> [The Engine] might act upon other things besides *number*, were objects found whose mutual fundamental relations could be expressed by those of the abstract science of operations, and which should be also susceptible of adaptations to the action of the operating notation and mechanism of the engine. Supposing, for instance, that the fundamental relations of pitched sounds in the science of harmony and of musical composition were susceptible of such expression and adaptations, the engine might compose elaborate and scientific pieces of music of any degree of complexity or extent.

While Babbage's machine was never built, Byron came to be celebrated for acute insights into its possibilities that foreshadowed developments in computer science a century after her death in 1852. She articulated a lucid explanation of how a machine capable of working with numbers could be used to process any kind of information or symbols that followed a set of rules. This idea would be the key to taking machines from calculating to computing, from number crunching to information processing.

In the final section of her notes on the prospective Analytical Engine, Ada Bryon included a detailed plan for using the machine to calculate Bernoulli numbers, an advanced method of mathematical analysis. With no term yet for what she did, nor a working machine to use it, Byron essentially wrote a computer algorithm.

Among the computers actually built in the mid-twentieth century, some of the first were shown to be capable of making music, if only for novelty effect unrelated to their primary purpose. As early as spring 1949, the programmer Betty Snyder (formal first name Frances, later married surname Holberton) was working on a new computer system, the BINAC (Binary Automatic Computer), and ended up using it briefly, at least once, to make a bit of music. Snyder had already served as one of the original programmers of the ENIAC when it was still a military secret and programming was a new, unfamiliar field conceived vaguely, naively, demeaningly as something akin to a clerical task: women's work. A wizard at math and science, Snyder was a trailblazing figure in early computing and created the first instructional code for the UNIVAC, the first general-purpose computer. (After that, she would help create the COBOL and Fortran languages employed in the world's most powerful computers of the time.) When the BINAC was fully programmed and ready for use, the staff gathered to celebrate, and Snyder treated the assembled to the sound of the computer "singing songs and stuff and playing music," including a rendition of "For He's a Jolly Good Fellow" in synthesized tones. "The performance was quite a hit," she would recall in later years. Not quite the premiere of *The Rite of Spring*, the occasion was a startling accomplishment of a new kind and a moment of minor history.

By the summer of 1956, at least a dozen scientists, musicians, and others experimentally inclined—and probably more whose work was undocumented or is now forgotten—had devised methods to create or synthesize music with some version of computer technology, in addition to Lejaren Hiller and Leonard Isaacson. In 1951 in Manchester, England, a math and physics instructor named Christopher Starchey was helping to program the Mark II, a general-purpose computer built on the principles Alan Turing had introduced, and he hatched a way to have the machine play "God Save the King." (Starchey also tried the Glenn Miller swing number "In the Mood," but that caused the Mark II to crash.) Writing in *Scientific American* in 1956, the engineer Richard C. Pinkerton described having designed a "banal tune-maker" to write singsongy nursery-rhyme melodies through binary selection, essentially flipping electrical coins. At RCA, meanwhile, a pair of electrical engineers, Harry Olson and Herbert Belar, were under a directive to research the sonic components of top-selling records so as to make hit-making more predictable and reproduceable, and got sidetracked by experiments in music synthesis, eventually constructing a room-size computer for replicating sound, which they called an Electronic Music Synthesizer.

"The liberal arts should not shrug off advances in science and technology as too technical to understand, and engineers at their end should not regard music and the arts as outside their natural domain," proclaimed David Sarnoff, head of RCA, in announcing development of its technology. "For more than a quarter of a century, the entertainment arts have felt the magic touch of electronics. As a result, music, drama, motion pictures, the phonograph, and even journalism have taken on new dimensions."

With the exception of Pinkerton's exercise in admitted banality, all these efforts of the postwar era focused on using the emerging computer technologies to play music, rather than create it: music-making as interpretation, not a generative act. Indeed, it was the proposition that the Illiac could not merely play a work of music but compose one that made headline news and stirred such unease among people who treasured humankind's capacity for creativity as one of the things that make people human. The fear of machines overtaking humans had run deep since the rise of the Industrial Age and typically took the form of distress over the dehumanizing regimentation of industrial labor: Henry Ford's one-worker, one-task philosophy as the nightmare image of Charles Chaplin's assembly-line worker, stuck in the repetitive motion of twisting a lug wrench after the lunch whistle has blown in *Modern Times*. But the particular fear of machines making art runs especially deep, for cutting into one of the human species' few claims to special status in a universe where ever-growing scientific knowledge makes us feel ever smaller.

There was a cartoon published in Europe in the era of John Nevil Maskelyne's automata. Rendered in strong, quick-gesture line drawings, it depicted an orchestra in the midst of a performance. The musicians had mechanical bodies—rods and pipes for legs and arms, bolted at the joints—with freakish human heads jutting out of metal-tube necks. They grinned with expressions of scowling menace as they played an assortment of instruments: violin, cello, organ, bassoon, French horn, trombone. A bass drum played itself with spindly wire arms attached to its sides. The conductor was a grandfather clock with a trap door on top, out of which protruded a thin mechanical arm with a human hand waving a baton. The atmosphere was one of free-flying monstrousness, a musical parallel to the images of crossed-species creatures put together in parts in H. G. Wells's *Island of Dr. Moreau*. Under the art was the caption: "A Prophecy." The message was clear in scratchy, angry strokes of black and white: Machines are taking over. The image is more scary than funny.

In the same summer as the premiere of *The Illiac Suite*, a weekly TV documentary series shot in California, *Adventure Tomorrow*, broadcast an episode about an experiment conducted that month by ElectroData, a Pasadena-based computer company that had just been purchased by the Burroughs Corporation, a major manufacturer of adding machines and office equipment. The host of the show, Martin Klein, was a telegenic, bespectacled aerospace engineer and part-time professor at Pierce Junior College, as well as co-creator of the project that was the subject of this episode: a computer program to compose music.

A collaboration with another engineer, Douglas Bolitho, the composing program utilized a new computer, the Datatron 205, which Burroughs was now offering for sale at a price of $135,000, not including the control panel for operating it and input/output gear for the punch cards. It was an impressive-looking machine, with blinking lights and spinning reels of tape. (The Datatron 205 represented the computer science of its time so well that it would be used as the computer in the Batcave of the *Batman* TV series of the 1960s, and would later appear in Dr. Evil's lair in the *Austin Powers* films.) Klein and Bolitho, under commission from ElectroData/Burroughs, analyzed one hundred popular songs of 1956, applied some standard guidelines for composition, and made a list of six rules for the program to follow, such as:

Do not skip more than six successive notes in the scale.

The first note in the first section of the song should not be the second, fourth, or flatted fifth note in a scale.

Notes altered with a flat should be followed by notes a tone lower, while those altered with a sharp should be followed by notes a tone higher.

Klein and Bolitho then used the Monte Carlo method to have the Datatron apply these rules to notes the Datatron 205 generated randomly. As Klein explained in an article for *Radio Electronics*, a magazine for kit-assembly hobbyists, he and Bolitho had "set out to prove that if human beings could write 'popular music' of poor quality at a rate of a song an hour, we could write it just as bad with a computing machine, but faster."

According to Klein, the program could generate one thousand songs per hour. On *Adventure Tomorrow*, he offered evidence of one, a song titled "Push Button Bertha" after the programmers' nickname for the Datatron 205. Burroughs hired Jack Owens, a journeyman songwriter and radio singer who had a bit of success writing geographical novelties such as the Swedish-themed "The Hut-Sut Song" and Hawaiian-themed "I'll Weave a Lei of Stars for You," to add words to one of the melodies the computer generated. Owens himself warbled it on the air.

The tune was perfectly adequate generic pop in the traditional mode—no better but no worse than many of the songs on the Hit Parade in the mid-Fifties, notwithstanding numbers in the new rock 'n' roll style that were beginning to make the Hit Parade obsolete. The lyrics added by the human songwriter for hire, Owens, were suitably clever:

> Sweet machine, what a queen
> Calculatin', palpitatin' chick with a click . . .

Widespread coverage of "Push Button Bertha" tapped into and fueled early computer-era fears of the incursion of computers onto human territory, while needling the commercial music industry for having revealed the earmarks of industrialization. "The report of a new electronic brain that can write 1,000 Tin Pan Alley songs in a single hour comes as a shock," wrote a reporter for the *Hartford Courant*. "Most of us thought such a machine had long been in existence."

Papers around the country stoked long-simmering anxieties over automation. "If the Datatron, by arranging musical notes in formalized sequences, lives up to its promise, lovers of real music can view it with as much suspicion as they would an atomic bomb stuck inside a radio or TV set," wrote a reporter for the *Austin Statesman*. "It has a potential for destruction of our remaining aesthetic sense in music quite as devastating as a 14-megaton detonation."

The *Michigan Chronicle* warned readers, "If you should happen to be walking in the vicinity of Tin Pan Alley's Brill Building in New York, be sure to wear a steel helmet and carry an umbrella. The purpose of these is to ward off the bodies of composers who might take the easy way out by throwing themselves from an upper story window. The suicidal tendencies are brought about by the news that songwriters (ever a sensitive lot) are in danger of being replaced . . . by a machine!" A computer had replaced human-headed robot monsters, but the cartoon narrative of music-making machines doing monstrous things was more alive and no funnier than ever.

7

BRAIN AUTOMATION

The family tree of automata split into widely divergent branches over the years following the heyday of Zoe the art-making automation in nineteenth-century London. Autonomous mechanical beings became fixtures in science fiction literature, films, comics, and TV shows, from the girlie stainless steel pinup in Fritz Lang's utopian dreamscape, *Metropolis*, to the robot-man servants in movie parables of the Atomic Age like *The Day the Earth Stood Still* and *Forbidden Planet*. By the time of the latter film's release in 1956, it seemed a given of scientific progress that humans (and aliens who looked just like them) would someday create humanoid machines with physical powers, advanced weaponry, and intelligence to surpass the capabilities of the human species. *Forbidden Planet's* Robby the Robot could speak 188 languages "along with their various dialects and sub-tongues" and crumble a ray gun with his bare metal hands. In fact, the monster in the climax of the film's plot was a gigantic computer brain, the out-of-control electronic id that Robby's master, Dr. Morbius—the Prospero figure in this drive-in update of *The Tempest*—built to run everything on his planet. To evoke the sci-fi horror of computerized intellect gone wild, *Forbidden Planet* had the first all-electronic movie score, composed and performed on homemade gear by the married collaborators Bebe Barron and Louis Barron, based on principles they found in a book by the MIT mathematician Norbert Wiener, *Cybernetics: Or, Control and Communication in the Animal and the Machine.*

In nonfiction science, meanwhile, the concept of automata evolved to become the intellectual framework for advanced thinking at the intersection of physics, math, linguistics, computation, and psychology during the postwar years. In 1956, the same year *Forbidden Planet* was released—and *The Illiac Suite* and "Push Button Bertha" made their premieres—a collection of new scientific papers and essays was published, titled for the name of its discipline, *Automata Studies.* The editors who conceived the book and commissioned its contents were two young mathematicians who had met while doing summer work at Bell Telephone Laboratories, the commercial research center with a program of public service in New Jersey: John McCarthy, who had joined the faculty at Dartmouth a year earlier, and Claude Shannon, who had recently started teaching at MIT. The core issue in automata studies, as McCarthy and Shannon saw it, was the potential of machines to think like human beings. It was a matter McCarthy had been absorbed with for several years, ever since he had attended an academic symposium on "Cerebral Mechanisms" at the California Institute of Technology. In his early work in this area, he conceived of "interacting finite automata" in mathematical form—not moving and drawing or trumpet-playing mechanical people, but conceptual automata expressed as math, akin to algorithms—to perform what he called "brain automation." While participants at the Caltech symposium had drawn eye-opening comparisons between the brain and computers, "No one had talked about actually using computers to behave intelligently," said McCarthy.

McCarthy was unhappy with most of the submissions he and Shannon received for *Automata Studies*, because they were nearly all about automata in various conceptual permutations but barely touched on the potential of machines to learn. "I was hoping that people would write about artificial intelligence, and one or two did," McCarthy would later recall. "But a number of them wrote just about automata." Among the problems he faced in getting others to think about artificial intelligence is the fact that no one had come up with that term for it yet.

"I didn't have the phrase," McCarthy said, until he set out to assemble a group of sharp minds from related disciplines to discuss the topic and realized he needed to be clear about what they would be discussing. "The idea was that people would come for the summer and work on artificial intelligence, and I had to call it something. So I called it artificial intelligence."

McCarthy's dream was to have something like a Manhattan Project for a field of science so new he had to name it. "My idea was to have a summer study analogous to the summer studies that had been held around some defense problems," he said. "I thought if one could get a lot of smart people, one could make substantial progress in a summer. That turned out to be very much overoptimistic."

Working in collaboration with three other scientists, McCarthy petitioned the Rockefeller Foundation to fund what their proposal described as "a 2 month, 10 man study of artificial intelligence" during the summer of 1956 at Dartmouth. Along with McCarthy, the organizers were Shannon of Bell Labs, the coeditor of *Automata Studies*; Marvin Minsky of Harvard, whose PhD dissertation on "Neural Nets and the Brain Model Problem" would be recognized as one of the founding documents of AI research; and Nathaniel Rochester of IBM, who had been experimenting with computers to test theories in neurophysiology. Their objectives, as they laid them out for the Rockefeller Foundation, grew from radical propositions they downplayed in dry prose:

The study is to proceed on the basis of the conjecture that every aspect of learning or any other feature of intelligence can in principle be so precisely described that a machine can be made to simulate it. An attempt will be made to find how to make machines use language, form abstractions and concepts, solve kinds of problems now reserved for humans, and improve themselves.

The speeds and memory capacities of present computers may be insufficient to simulate many of the higher functions of the human brain, but the major obstacle is not lack of machine capacity, but our inability to write programs taking full advantage of what we have.

With funding granted by the Rockefeller Foundation (at roughly half the sum McCarthy and his partners had requested), the Dartmouth Summer Workshop on Artificial Intelligence—the first project of any kind to use that term—began on June 18, 1956. In one suitably quantifiable, purely numerical sense, the workshop was at least twice as successful as its organizers had envisioned. More than twenty people, rather than the projected ten, attended at one point or another, however briefly, and they matched the profile of the proposal: all of them were men. Twenty white men. The sole known exception was a pediatric fellow from Boston, Gloria Minsky, who accompanied her husband, Marvin. Over the course of the workshop—accounts of its duration vary from six weeks to seven or eight—people came in and out. Granted open access to the math department space on the top floor of Dartmouth Hall, the handful of people in attendance at any one point would gather in one of the classrooms and talk. There were some formal and almost formal presentations on some of the attendees' research, with abundant time for informal chats on matters ranging from heuristic logic and name theory to the relationship between the bandwidth of television broadcasts and the human eye's ability to infer abstractions. The mathematician Trenchard More came for three weeks and for much (but not all) of that time took notes that captured the freeform casualness of the workshop:

> Culver accompanied McCulloch when they came to take Ashby to a conference at Colby College. Culver is interested not in artificial intelligence, but inertial navigation.
>
> McCarthy has become increasingly interested in the problem of writing a program that will write programs, the generative program being part of a general problem solving program.
>
> McCulloch was present only for a few hours. In that time, however, he discussed Carnap's system of languages, and Turing machines. . . . McCulloch claims that he can prove that a Turing machine can learn the English language, and believes that the human brain is a Turing machine.

Tapping the historical traditions of formal logic, factoring in philosophical concepts of human thought, and applying voguish midcentury ideas about systems engineering, the people gathered for the Dartmouth Summer Workshop on Artificial Intelligence proceeded from the belief (or hope) that the workings of the mind could be deduced from physiological functions of the brain. Once they were mapped out as a system, those functions could, hypothetically, be replicated electronically by computer. In establishing this framework, the workshop did something more important than introducing the term "artificial intelligence"—it gave the new field of study a conceptual structure, a way for adventurous thinkers to think about it and conduct future work, at least for a while. That computers might have the capacity to provide unique models for intelligence and methods of achieving it that could rival or surpass those of human intelligence was certainly unthinkable in 1956, even among those doing the most advanced thinking about how both humans and computers think. For McCarthy and his colleagues, emulation of human intelligence was a sufficiently challenging objective, without worrying about the theoretical limits of humankind.

By all accounts, the only three people to stay for every day of the workshop were McCarthy, Minsky, and the mathematician Ray Solomonoff, who presented and handed out hard copies of a paper outlining a plan for what he called "An Inductive Inference Machine." It was the first known articulation of algorithmic probability and would come to be recognized as one of the foundational documents of machine learning,

In the end, John McCarthy was unsatisfied. "Certain things didn't happen as planned," recalled McCarthy, a burly man with an imposing affect of focus and purpose. There is a photo of seven of the men who attended the conference assembled under a tree in front of Dartmouth Hall. (See page 92.) Six are sitting or lying on the lawn, laughing and grinning, while McCarthy squats down alongside them without letting his behind touch the ground, like a baseball catcher, half-smiling for the camera. "Most of the people who came could only come for a short time. I thought we would make substantial progress toward human-level artificial intelligence that summer, and we didn't.

"Anybody who was there was pretty stubborn about pursuing the ideas that he had before he came," said McCarthy. "Nor was there, as far as I could see, any real exchange of ideas. People came for different periods of time. The idea was that everyone would agree to come for six weeks, and the people came for periods ranging from two days to the whole six weeks, so not everybody was there at once. It was a great disappointment to me because it really meant that we couldn't have regular meetings."

Clearly, the notion to stimulate free, cooperative collaboration by putting a group of twenty white males together in a building for six weeks during the summer was not John McCarthy's soundest conjecture. Protective of the disparate ideas they were developing independently, perhaps tentative about sharing them or fearful of having them pilfered, most of the participants at the Dartmouth Workshop sprayed their intellectual territory and left. Somehow, there had seemed to be more productivity at the gatherings for military research that had been John McCarthy's model, where the purpose of bringing groups of men together was to develop methods to make big explosions and threaten people with destruction.

THREE HUNDRED YEARS AFTER THE DEATH OF RENÉ DESCARTES, THE VIEW OF humankind as machinery endured in postwar America and, in fact, only grew in cultural resonance. The character of machinery had changed sig-

nificantly, of course, with clockwork and gearing now supplanted by computers and electronics in the popular imagination. (In daily life, traditional machinery did not suddenly disappear and would maintain a presence in every kitchen and toolshed into the twenty-first century.) That change, in turn, began to transform the way human beings conceived of themselves. Where the machine model had once helped us understand the human body (our muscles and bones, blood flow and heartbeat) in terms of Newtonian physics, a new category of machines led us to imagine the brain (how we think, what we know, even how we feel or how we think about what we feel) in terms of the computer. If information could be thought of as data, now, too, could thought itself.

As John McCarthy (see photo, page 98) and Claude Shannon wrote in the preface to their book *Automata Studies*, "Descartes, in *De Homine* [*Treatise on Man*], sees the lower animals and, in many of his functions, man as automata. Early in the [20th] century, when the automatic telephone system was introduced, the nervous system was often likened to a vast telephone exchange with automatic switching equipment directing the flow of sensory and motor data. Currently it is fashionable to compare the brain with large scale electronic computers."

They could have added Sigmund Freud after Descartes in that history, for his description of the mind as working like the most advanced technology of his day, the steam engine. When pressure builds up in the psyche of a person or the boiler of an engine, it needs to be released. Freud considered various options for the form such a release could take in a person, including acting out at the perceived source of the pressure: blowing off steam.

In the case of computers and the brain, the basic parallels are clear, and so basic as to be both irrefutable and also not much use after a time. From the conceptual prototype of all computer technology, the Turing machine, to the complex linked networks of the twenty-first century, all computers have essentially done the same work: They have taken input, processed it, and produced results. Among the profound changes wrought by high-level machine learning is the ability of machines to find their own input and make of it what they will. Still, the main work they're doing is taking input, processing it, and producing results, and that is what the human brain does. In the words of Allen Newell and Herbert Simon, two scientists who attended the Dartmouth Summer Workshop, the computer and the brain are "both species belonging to the genus IPS" (information problem solving). On the level of elemental function, then, the brain is clearly analogous to a computer, half of a useful metaphor in the tradition of the automata that inspired Descartes or the steam engine in Freud's day.

John von Neumann, a key inventor of computer architecture and lover of march music, gave the first of an intended series of lectures at Yale not long before he fell terminally ill and had to curtail his work. The topic was the relationship between the computer and the brain, which he described as "two kinds of automata." (His death from cancer at age fifty-three in 1957 has been correlated to his exposure to radiation at Los Alamos during his Manhattan Project work.) "The most immediate observation regarding the nervous system is that its functioning is *prima facie* digital," von Neumann said, as the lecture text was posthumously published in book form. As neuroscience had already demonstrated, the workings of cellular activity in the brain have some literal parallels to digital processing, as well as considerable differences that von Neumann could only surmise in the 1950s. Neural function qualifies as binary, in that a neuron will fire or it won't; it's either on or off, tantamount to values of zero or one. At the same time, a range of subtle, variable phenomena are taking place with each neuron, rendering its status as on or off just one aspect of its character and function at the time. Indeed, neuroscientists still don't know what all those variable things are, let alone how they work in relation to one another.

Every neuron is in constant flux, influenced by synaptic activity and neuromodulators, and each one is only one acting in some way with billions of other neurons interconnected in a gargantuanly complex web. No—the image of a web is too neat and formal, too graspable to apply here. We have no comprehension of how the interconnections of neurons would look to us if we could freeze them for an instant, and stopping them for a photo would misrepresent the way they work, anyway.

Early computers were designed with the goal of emulating human intelligence. To conceive of the brain as a computer is to think of it as a mirror of the thing doing that thinking, and it's a trick mirror, reflecting its object only in part. Computers were invented primarily to replicate only one kind of human cognition: computation. As every conscious human knows, doing computations—whether with number or symbolic values, consciously or unconsciously—is just one aspect of consciousness. Among the more elusive matters of consciousness is "the problem of what understanding is at all," in the words of Joseph Weizenbaum, a computer scientist at MIT. "The A.I. notion is that understanding something—for example, *Othello* or *Tristan and Isolde*—is to be able to say what events transpired as the drama unfolded. But it's a far from sufficient proof of understanding to merely be able to say, 'these events transpired.' There's a missing piece here which I would say is fundamentally technology-independent."

For computers to do what the brain does, they have to do a great deal more than process information with magnificent speed and sophistication. They need to comprehend as well as act; they need to question as well as answer, to learn as well as know; they need to feel as well as think; and they need to be capable of producing many things other than knowledge. They need to have the aesthetic intelligence and emotional capacity to appreciate art, and the creativity, if not the *spiritus*, to make it.

8

EVERYBODY SHOULD BE A MACHINE

No relation to the pornographer of a later day, who spelled his surname with a "y" for the vowel, Larry Flint was an expatriate American artist living in Paris during the early 1960s, a variation on the expat bohemian cliché of a character played by Gene Kelly a decade earlier in the MGM musical *An American in Paris*. Flint, bearded and poor, was an ascetic purist inclined to spout idealistic bromides like "Money corrupts! Art erupts!" He was enraptured with the art of a painting chimpanzee who worked in his *arrondissement*, and once praised a canvas of the chimp's bright-colored splotches as "a testament to the human spirit—total primitive archetype!"

Flint's own work was made by means no more orthodox than those of a painting chimpanzee, and its spirit was no more human. Flint invented a machine that harnessed sound waves to generate mechanical movements through a jerry-rigged setup of electronics, levers, pulleys, and metal arms with paintbrushes attached at the end. The concoction responded to the mood of the ambient sound, selecting colors and applying paint in appropriate gestures to create artworks with a corresponding visual aesthetic. Flint, bounding manically around his garret, would bang a hammer on an anvil, ring a fire bell, pat out a rhythm on a set of bongo drums, and run a jackhammer, and the machine would produce fittingly random, visually cacophonous art. "The sound, the sonic vibrations, they go in there, and then they get transmitted to that photo-electric cell, which gives those dynamic impulses to the brushes—it's a fusion of the mechanized world and the human soul! That's the only affirmative statement being made in the world of art today," Flint explained to his wife, Louisa, who influenced him significantly by suggesting that he play records of Mendelssohn for the machine so it would use the pretty sounds to make art of more conventional beauty.

Mechanical devices of many kinds had been employed in the making of visual art for centuries by this time. The pantograph, a mechanism of hinged and jointed rods, had been used for altering the scale of things for drawing since the first decade of the seventeenth century. Plotters such as the perspectograph, a machine for coordinating the eye and the hand in relation to objects, were in use by the seventeenth century, too. The camera obscura—a pinhole in a wall or side of a box that exploited a quirk of optics to project images from life onto surfaces or systems of mirrors for tracing—was known to have been employed as early as 500 BCE and flourished during the Renaissance, when the revered masters of Western art history had assistants using machines.

Larry Flint died at the hand of his own machinery, when the art it erupted forth brought him too much money that was too-too corruptive and his greed made the equipment go haywire. It was a predictable and not particularly funny resolution to a strange, aggressively parodic morality tale, one in a series of allegorical vignettes in the Hollywood musical fantasy *What a Way to Go!* from 1964. Written by Betty Comden and Adolph Green, who had gotten their start doing topical satire in a Greenwich Village revue act before writing *On the Town* and *Singin' in the Rain* made them famous, *What a Way to Go!* told the story of an accidental black widow (played by Shirley MacLaine), whose homey aspirations to true love and the simple life inadvertently make a string or husbands rich and, one after another, dead. (The victims were played by a gallery of male movie stars of the Sixties: Paul Newman as Larry Flint,

Gene Kelly, Dean Martin, Robert Mitchum, and Dick Van Dyke.) With the story of Flint and his wild painting machinery, Comden and Green took on the avant-garde world for the first time since *The Band Wagon*, their knife-twisting homage to the experimental New York theater scene of the early postwar years, this time using art-making machinery as a symbol for pretentiousness in the same way they had used *Faust* in the earlier film.

Three years before Paul Newman appeared in *What a Way to Go!*, he had portrayed another expat American artist—a jazz musician seeking camaraderie in the aesthetically and socially adventurous atmosphere of Montmartre—in Martin Ritt's film *Paris Blues*. The Larry Flint character, comparable but bearded, had ideas even wilder than a dream that jazz could be recognized as serious music, and his susceptibility to some provocative notions about art made him a comically topical kind of aesthetic radical. He believed a chimp could make art of the human spirit, and he also loved the work of a street artist whom he watched as she shot guns at paint-filled balloons hanging in front of an enormous canvas. Animal art! Violence as art! Art by machine! *A fusion of the mechanized world and the human soul.*

For *outré* art-making by machine to be the target of satire in a big-budget Hollywood musical—even a peculiar, not very good one like *What a Way to Go!*—there would have to be a fair number of moviegoers who would find the idea vaguely plausible and at the same time comical: the sort of thing they might see in the back pages of *Time* magazine and chuckle over as the latest wackiness from the ever-wacky art world. Andy Warhol would have primed them.

By this point in the Sixties, Warhol was becoming well known for his ready-for-ridicule experiments with mechanical production, commercial products, and their iconography, making news with his large-scale canvases depicting the purest symbol of industrial capitalism, dollar bills; his paintings of brand-name, mass-market goods like Campbell's soup cans and Coca-Cola bottles; and his reverently meticulous, three-dimensional reproductions of Brillo boxes and packages for Campbell's soup and tomato juice. (See photo, page 102.) He was beginning to be recognized—and, in some corners, bemoaned—for disrupting traditional lines of delineation between machine work and the work of human hands, between commercial and fine art, high and low, entertainment and art, fun and seriousness.

A former commercial artist, Warhol came from a place where machine methods and technology were broadly applied and widely accepted. Artists doing advertising, marketing, and industrial work routinely used photographs—images made by cameras, picture-taking machines, and enlarged and processed by machines—sometimes as the primary imagery or in design constructions, other times as source material, traced over with the photos on a lightbox. Commercial art was created and disseminated with machines involved in many ways, from making color separations for reproduction (a process that allowed for myriad adjustments while introducing its own quirks and artifacts), screening (through the use of Ben Day dots, which Roy Lichtenstein exaggerated in paintings inspired by comic book panels), and various methods of printing. TV commercials had stop-action Alka-Seltzer tablets singing songs and Hertz rental car customers flying from the sky into their drivers' seats, all delivered through the airwaves on staticky, flickering TV screens. No one could fault commercial art for not being handcrafted or relying on printing or broadcast technology for its delivery; commercial art was inextricable from technology.

Warhol had been working from photos and applying industrial-art techniques since his days as an illustrator for shoe catalogs and magazine ads. To make his large-scale grids of paper currency, *200 One Dollar Bills*, he first rendered a sample bill by hand—Warhol could draw—and reproduced it 200 times by silk-screening. The process, though applied to fine-art purposes from time to time, was most familiar in the U.S. from its commercial applications. Fast, efficient, and economical, silk-screening was widely used on ads and promo materials made for quick turnaround and imminent replacement, such as supermarket flyers. (For variations of currency art—*32 One Dollar Bills, 15 One Dollar Bills*, and more—Warhol used a variety of printing methods, including rubber stamps he had made of the face of bills.)

"Warhol's currency paintings (he made lots) gave his public its first encounter with the famous idea that he wanted to function like a machine—in this case, a printing press in the U.S. Mint—and that his art had nothing to do with the touch of his hand or his manual skills and labor," wrote Blake Gopnik in his biography, *Warhol*.

The idea Gopnik refers to was still as infamous as famous in 1964, when the full complexity of Warhol's aesthetic was not yet clear. "Everybody should be a machine," Warhol told the critic Gene Swenson in a Q & A for *ARTnews* in November 1963. In the next sentence of the published interview, Warhol added, "I think everybody should like everybody."

Swenson followed up: "Is that what Pop Art is all about?"

"Yes," answered Warhol, "it's liking things."

Pressing for clarification on Warhol's aesthetic philosophy, Swenson asked, "And liking things is like being a machine?"

"Yes," Warhol replied, "because you do the same thing every time. You do it over and over again."

Decades after the interview was published, the scholar Jennifer Sichel found and listened closely to the original tape of Swenson's interview, and she learned that Warhol was actually offering a more layered argument to make a point about liking between people, including people of the same sex. In a question cut for the published version of the interview, Swenson asked, "You mean you should like both men and women?"

Warhol answered, "Yeah."

Swenson continued, "Yeah? Sexually and in every other way?"

"Yeah," said Warhol.

"And that's what Pop Art's about?"

"Yeah, it's liking things."

Apart from the minor transgression of changing Warhol's casual yeahs to stilted yeses, Swenson erased what, in 1963, was just as bold, if not more bold a public statement than the assertion that consistency in personal taste can be thought of as mechanical. Warhol's original comment saying that everyone should be a machine was a call for people to be equitable in their enthusiasms—not dispassionate but democratic in their passions. The point was not simply to like things, but to like all things with open appreciation. The parallel to machinery lay in the inability of inert machines to pass judgment, as well as the capacity for machines to engage in endless acts of repetition without pause for reflection or change of heart.

Warhol spoke a tiny bit more about his attraction to mechanical methods elsewhere in his interview with Swenson. "Mechanical means are today, and using them I can get more art to more people," Warhol said. "Art should be for everyone." He understood the power of popularity in a culture driven by celebrity and fashion, and he wanted, he *needed* to be trendy, especially if he could invent the trend himself. He sought the means that were "today"—or better still, *right now*, this instant in the flash of a Polaroid flashbulb. With the popular appeal that trendsetting conferred, Warhol could and did make art for everyone: mass-produced portraits of entertainment stars like Elizabeth Taylor, Marilyn Monroe, and Elvis Presley, art that was likable and accessible, but with complicating elements of doubt, mystery, and sadness in the non-accidental technical imperfections that stain their glossy surfaces.

As Gopnik described Warhol's first silk-screen portrait, one of Elizabeth Taylor, "This crude transfer onto fine canvas of a cheap photo of Liz turned the painted face of a human being, with its potent evocations and links to Great Art, into something as 'blank, blunt, bleak, stark' as any consumable product—which was precisely what Hollywood's media machine had made of Liz. Or maybe the relevant comparison is even blunter: Liz sits on her canvas just as the hand-drawn greenbacks had in Warhol's first silkscreens, the actress now become a perfectly fungible commodity." (With "blank, blunt, bleak, stark," Gopnik referenced a phrase Warhol's friend and art dealer Ivan Karp had used to describe Warhol's earlier paintings.)

A savvy role-player who exuded qualities of unguarded, almost childlike naïveté and vulnerability, Warhol would invariably talk about his processes in deceptively one-dimensional terms. Describing his embrace of silk-screening in his philosophical memoir, *Popism*, he wrote, "The rubber-stamp method I'd been using to repeat images suddenly seemed too homemade; I wanted something stronger that gave more of an assembly-line effect."

Handy simplification of this sort was of a piece with the simplistic criticism of Warhol's methods and their results as cold, impersonal, inhuman, and, worse, cynically anti-human. In 1964, the French writer Gérald Gassiot-Talabot gave voice to the early dissent on Warhol among critics who saw the human hand and the human heart as exclusive to humanistic art. "Warhol and his peers demand that we radically revisit the criteria for aesthetic appreciation—insofar as these artists assert, with often delicious cynicism, a conceptual and technical laziness whereby they prefer the methods of automatic reproduction to the vagaries and tiresome aspects of traditional craft," wrote Gassiot-Talabot in an essay on Warhol and the young Pop Art movement for *Art International*. "Unfortunately, so long as the definition of art presupposes personal, voluntary intervention by the painter, Warhol will reside in that indistinct fringe of creative depersonalization where the refuse collectors of Nouveau Realisme have erected a whole swath of current artistic practice, while possessing a human note that Pop does not."

Others with more appreciation for Warhol's processes and output have argued, by contrast, that the assembly-line methods Warhol liked to talk about were deceptively, superficially mechanized and belie the fastidious handiwork of an activist creator. In Gopnik's words, "If silk-screening always evoked its 'low,' industrial roots, that was in fact at odds with how Warhol used it as a fine art technique, as hands-on and small-scale as could be."

It's best to assume that Gopnik meant not to suggest that Warhol's technique was as hands-on and small-scale as *any* art could be. Let's take the claim to be a relative one, an irrefutable observation that Warhol's technique was as hands-on and small-scale as *it* could be. Or: as hands-on and small-scale as work could be that was made by Warhol with various assistants, sometimes working on their own without Warhol present, knowing or inferring Warhol's intentions. This way of thinking is scarcely novel or scandalous; the long history of celebrated artworks from the Sistine Chapel and the *Pietà* to *Guernica* to the stainless steel balloon dogs of Jeff Koons is a history of collaboration among artists and their assistants, some (or many) of whom have often proceeded with intermittent (or less) supervision.

The more telling issue in the case of Warhol is not how significantly his assistants might have contributed to the character of art made through mechanical processes, but how significantly the *machines* contributed to that character. We can use the example of the silk-screen portraits of Elizabeth Taylor. The works are powerfully expressive. "For me," the critic

Jerry Saltz wrote, "Elizabeth Taylor is Andy Warhol's paintings of her. When I see her, I see those amazing pictures, and my mind is flooded with ideas of image manipulation, overexposure, divadom, iconicity, victimhood, vampire, parody, celebrity, cult heroes, lost soul, loving freakiness, queerness and fearlessness."

Warhol made more than fifty Taylor portraits and each one is unique, distinguished by variations in the colors of her face and her makeup. Apparent glitches and flaws appear and disappear from one portrait to the next: the lipstick is smeared; the eyeliner is covering part of her eyes; the hairline is off-register, like a cartoon image in a cheap old comic book. Of course, the glitches cannot really be flaws by definition, because they're part of Warhol's art—elements of its identity, not detractions from that identity. Take away the myriad aspects of human involvement—a slip of the screening squeegee, a thick glob of paint obscuring a small detail on the publicity photo the portrait is made from—and what's left is not a Warhol. If every aspect of the silk-screen were perfectly photorealistic, the result would be a pristine and banal sample of commercial representation, rather than an inversion or reinvention, a commentary on such a thing.

Then again, take away the silk-screen process—the way it reduces the original photograph to a chiaroscuro blueprint; the way it applies colors in crude, flat layers; the way its tight calibrations encourage mechanical error—take away the machine, and all that's left is that pristine, banal commercial photograph. Without Warhol, the silk-screen portrait of Elizabeth Taylor is not fine art; and without the silk-screen machinery, it isn't a Warhol. It is worth keeping in mind, too, that the process began with a PR photo Warhol did not take; a camera came first.

Warhol, invariably simplifying or obfuscating in his terse, often cryptic public statements, said he prized what the machinery brought to the art more than he valued human intervention. In late 1962 and early 1963, he sat for a series of interviews with the critic David Bourdon, whom he had known since his days as a commercial artist. Bourdon pointed out, "It's impossible to make an exact copy of any painting, even one of your own. The copyist can't help but contribute a new element, or a new emphasis, either manual or psychological."

Warhol replied, "That's why I've had to resort to silk screens, stencils and other kinds of automatic reproduction. And still the human element creeps in! A smudge here, a bad silk screening there, an unintended crop because I've run out of canvas—and suddenly someone accusing me of arty lay-out! I'm anti-smudge. It's too human. I'm for mechanical art. When I took up silk-screening, it was to more fully exploit the preconceived image through the commercial techniques of multiple reproduction."

When he learned that the filmmaker and photographer Anton Perich had built a painting machine, an enormous construction that rendered erratic images on large canvases—shades of Larry Flint's movie contraption—Warhol was stricken with envy. "I was so jealous," he wrote in his diary. "My dream. To have a machine that could paint while you're away."

In a Q & A–style interview published in 1964, Warhol discussed mechanical production with his friend and studio assistant Gerard Malanga, who handled some of the machinery used in the making of Warhol's artworks. "Would you like to replace human effort?" Malanga asked.

"Yes," Warhol answered.

"Why?

"Because human effort is too hard."

A bit later, Malanga asks, "How will you meet the challenge of automation?"

"By becoming part of it."

And, "What does the computer mean to you?"

"The computer is just another machine."

The interview was presented in the literary journal *Chelsea* as "Andy Warhol on Automation: An Interview with Gerard Malanga." A more precise title would have been "Andy Warhol as Automation," because his presence was a simulation. Gerard Malanga wrote every word: the questions presented in his own voice and the answers offered in Warhol's, with the consent of Warhol. Malanga, who knew how Warhol spoke, if not what he thought and felt, served as Warhol's human automaton.

Warhol and his company would take the idea of automating the artist further, treating Warhol as no more than an idea that could be transferred from his body to another, such as that of Alan Midgette, the near-lookalike whom Warhol sent out to portray him on a speaking tour in 1967, or, going further still, the robot Warhol planned to stand in for him in a one-man/no-man Broadway show, "An Evening with Andy Warhol," in the early 1980s. Though the script was never completed and the production never staged, the robot Warhol was built by a former Disney engineer, Alvaro Villa, and photographed for features in *Life*, *Time*, *People*, and *Popular Mechanics*. "Andy loved this idea," recalled Bob Colacello, former editor of Warhol's own magazine, *Interview*. "The idea was that the show, if it was successful in New York, could then also simultaneously be running in London, Los Angeles, Tokyo with cloned robots. And people would actually be able to ask questions of the robot, which would be programmed with a variety of answers. The whole thing was Warholian and so perfect." Abandoned for lack of funding, the Warhol robot show would remain a deliciously unlikely idea, perfectly Warholian.

In the day-to-day production of silk-screen art, Gerard Malanga was the de facto director of manufacturing at Warhol's base of ill-defined operations, the Factory. Its first iteration, in a floor-through loft on the fifth floor of 231 East 47th Street in midtown Manhattan, was known for a while as the Silver Factory, because Warhol had had the artist Billy Name coat all the surfaces in aluminum foil and silver spray paint. Gift-wrapped on the interior, like a package from Saks Fifth Avenue turned inside out, the Factory was, in its early years, a place of glittery pleasure and unlikely productivity. "There was a reason they called it the Factory," recalled John Cale, the musician who worked and played there with Lou Reed and their bandmates in the Velvet Underground, which led Warhol to design the cover for their first album and be credited as the record's producer. "Andy would be sitting there with a tape recorder, doing I don't know what. Another person would be off to the side, making a silk screen. Someone else would be shooting a film. And a group of people behind them would be having a little orgy."

Warhol's ballooning fame and mystique as a genius cipher drew a midway troupe of artists and artists-to-be, wannabes, curiosity seekers, hangers-on, and their lovers to the Factory. Warhol reveled in the company of drag queens and freethinkers of every stripe, and his laissez-faire detachment from whatever spun around him inevitably stirred up an atmosphere of arty bacchanalia. In some ways, nonetheless, the Factory remained true to its name. There were people who carried responsibilities for specialized tasks, such as Ronald Tavel, who wrote scripts for films that Warhol shot there and at other locations in underground New York; Edie Sedgwick, who provided glamour and a conduit to Manhattan society; and Ondine, who helped keep the amphetamines in supply. The output of product was impressive: from 1964 to the end of 1968, Warhol oversaw or facilitated the making of more than sixty films shot at the Factory or with Factory characters, including touchstones of avant-garde cinema such as *Empire*, an eight-hour static shot of the Empire State Building; *Chelsea Girls*—essentially two films projected simultaneously, side by side, a voyeuristic patchwork of scenes in the Chelsea Hotel; and *Blow Job*, a thirty-five-minute study of the face of a man as he receives oral sex, projected at slow speed; as well as such nonsense as *The Nude Restaurant*, in which we see a number of Factory regulars in G-strings milling about a Greenwich Village eatery.

"The technical capabilities of the camera set the parameters for the films," explains Steven Watson, author of *Factory Made: Warhol and the Sixties*. "When Warhol discovered how cumbersome the process of editing was, he rejected it. The unedited, one-shot, three-minute take became an inviolable building block. Whatever happened on screen was shaped by duration. The early sound films ran thirty-five minutes or seventy minutes—nothing in between.

"The assembly line was the model for film production: Shoot the film one day, send it out to be developed the next, and by the end of the week publicly screen it at Jonas Mekas's floating Film Cinematheque. The production schedule was exhausting and exhilarating."

Warhol called the films made under his name "experiments," and their hit-or-miss quality, the pervading feeling of "let's see what happens if we do this"—or "let's see what happens if we do *nothing*"—makes them feel as much like the output of a laboratory as a factory. Warhol, who grew up in Pittsburgh when it was still a steel town, had a fascination with the grim theatricality of heavy manufacturing. Seated at a table with the head of U.S. Steel at a White House dinner for Imelda Marcos in the early 1980s, Warhol offered up an idea to retrofit one of the company's foundries as a tourist attraction. "What you should do is put one of the unused buildings to use and make it into a Disney World and give tours and charge people $10 to get a little coal on their faces and see the hot lava being poured," he said, as he recounted the moment in his diaries. Still, the atmosphere at the Factory had little in common with the searing Pittsburgh steel mills and the roaring auto plants Iggy Pop was trying to mirror in music during the same period. "Iggy's concept of the mechanical was utterly American and industrial. Andy's was different," said John Cale, who knew them both. "Andy's concept was more advanced. Somehow, he had an understanding that machines could do things a human artist could not do."

© AMN 1965

GAUSSIAN−QUADRATIC (1963)
BY A. MICHAEL NOLL

9

PATTERNS

"There are no rules," the painter Helen Frankenthaler told an interviewer in 1994, when she was seventy-four and looking back on a career full of unexpected turns and experimentation. "That is what I say about every medium, every picture. . . . That is how art is born, how breakthroughs happen. Go against the rules or ignore the rules. That is what invention is about."

Well educated in art as a young woman, Frankenthaler had lessons in pictorial composition at Bennington College and, after graduating in 1949, studied privately with the painters Wallace Harrison and Hans Hofmann. She knew full well that there were such things as rules, formal and informal, in fine art: rules of perspective, principles of color theory and balance, standards of proportion, unity, and contrast. If there were no rules, there would be nothing to go against or ignore, no values to contest, no traditions to defy, no way for any artist's work—no matter how original—to be radically subversive and point to the norms of the future.

A fellow painter of the postwar era, Grace Hartigan, also followed the "no rules" rule, and railed against the oppressive pressure to conform to the male-dominated art vogues of her time. "I really loathe abstract expressionist painting down to the last splashy brush stroke," Hartigan wrote in her journal in 1952, when the cult of bravura action painting dominated the art establishment. "There is something about Matisse—" she continued, letting the thought hang, unresolved. "In the end though all these ideas mean nothing—you have to paint your way through it. No rules, I must be free to paint anything I feel."

In a realm where rules were tantamount to restrictions, Hartigan made gorgeous and vibrant, partly quasi-representational paintings that rejected the showy machismo of Jackson Pollock's splatter paintings, which in turn had represented a rejection of traditional brush technique. When Andy Warhol and the Pop artists came along, they would reject Abstract Expressionism in different ways. Donald Judd would go in yet another direction, contracting out his clean, smooth constructions to commercial manufacturers, while Alice Neal would ignore everyone and take up the classical art of portraiture while simultaneously disrupting it. In fine art, a field in perpetual flux, Frankenthaler and Hartigan captured the creative act with a single, self-contradictory command: No rules.

Computers didn't come to fine art until the 1960s, a full decade after they had been employed to make music, because computational technology relies upon rules. It thrives on rules. It abhors a rule vacuum. No rules, no way for a computer to do its magnificent data processing, and music of most kinds (but not all) has rules of mathematics at its core. Because the Western tempered scale is essentially a mathematical system, music lent itself neatly to early computers. (The tempered scale approximates the math of acoustic waveforms, but with compromises that give the math priority over the sound. Compared to the patterns of pure sound, the notes of the tempered scale are "off" at points.)

"Computers accept instructions and data by means of input mechanisms; they then do something with these data; and, finally, they return results to the user of the computer," wrote Lejaren Hiller. To train the mainframe at the University of Illinois to compose *The Illiac Suite*, Hiller and Leonard Isaacson began with a system for generating pseudo-random numbers (the Monte Carlo method) and coordinated it with the math underlying the rules of intonation, harmony, rhythm, and the like that had traditionally regulated Western classical music. "The method is justified in instances where the object of study can be assumed to be a statistically ordered universe from which we can isolate certain elements to form a simpler model universe characterized by events subject to the same statistical order," Hiller explained, sounding very much like the chemist he was when he wasn't being a composer.

Stravinsky, among other composers, appreciated the mathematical dimension in musical thought. As he told his late-life shadow, Robert Craft, music is "far closer to mathematics than to literature—not perhaps to mathematics itself, but certainly to something like mathematical thinking and mathematical relationships." Much the same, Thelonious Monk—a composer who broke musical time apart in minutely precise, deceptively erratic-sounding delineations—said that he thought "all musicians are subconsciously mathematicians." With the rise of the vogue for twelve-tone and serial music, modernist composers (including Stravinsky, after a time, following Arnold Schoenberg's lead) brought new levels of mathematical abstraction to classical music, essentially (though not fully) replacing traditional tonality with abstruse systems of reordered notes. Critics of this proudly challenging music faulted it for sounding mechanical and overly cerebral, connecting machines and the mind in opposition to the heart.

There is a danger in making too much of the math–music dynamic, however. Both mathematics and music are abstract forms, and both are concerned with measurable relationships between quantities; but music ultimately involves things not easily reducible to numbers, such as human feeling, and math is the key to much more than music. What's relevant here is the fact that music was companionable with early computers because classical music scores could be converted into numerical data, and certain principles of composition could be translated into programming instructions. There were rules in the music, calling out for computers to find them and apply them in a new manner.

When someone finally used the computer at Bell Laboratories to make a work of visual art, in the early 1960s, it happened by accident; and some people, including the person who made it, would not call it art.

In the spring of 1965, art collectors and patrons of the arts, critics, and others on the mailing list of the Howard Wise Gallery at 50 West 57th

Street received a cleverly unconventional invitation in their mailboxes: a business envelope containing a stack of four computer punch cards in different colors—yellow, blue, red, and green. It seems fair to presume that most people in the art world, like most people in the rest of the world then, had never before held punch cards in their hands and may not even have known what they were. Like their functional counterparts, the cards mailed out by the gallery had rectangular holes punched in cryptic patterns, as well as numerals printed in dense rows, along with an announcement of an upcoming show called "Computer-Generated Pictures." That show marked the first time any imagery produced by computers would be hung on the walls of any gallery or museum in the country, presented to art lovers the same way paintings and drawings were.

The gallerist, Howard Wise, was a fifty-seven-year-old newcomer to the clubby Manhattan art scene, having arrived five years earlier from Cleveland, where he had made a fortune in an inherited industrial paint and varnish business and begun a second career as an art dealer. Conversant in commercial technology from his years in the business of finishing products, he started to build a space age identity for his gallery a few months before the "Computer-Generated Pictures" show, by presenting an exhibition of plastic, 3-D optical art in moiré patterns—op art—made by a biochemist at the Polytechnic Institute of Brooklyn, Gerald Oster. As *The New Yorker* reported in an ogling "Talk of the Town" piece, Oster found himself curious about moiré patterns he had encountered in his research, while "his only thought was that they might make a useful laboratory tool" until "he noticed how beautiful the patterns were." This was aesthetic value rooted in science, noted the writer, Gerald Jonas: "The patterns are visualizations of mathematical relationships; their magic is the magic of math."

Now that the Oster show had successfully associated the Howard Wise Gallery with fresh and offbeat, technically oriented conceptions of art, Wise went scouting for other cases of inadvertent art-making outside the art world. "Wise was on the lookout for people doing unconventional things, and he appeared to have a particular interest in science," recalled A. Michael Noll, an engineer at Bell Labs at the time. Wise got in touch with a colleague of Noll's at Bell, Béla Julesz, a Hungarian émigré who was studying depth perception and had published an article in *Scientific*

American illustrated with graphics of dots in color patterns that Julesz had made at the labs. Wise asked Julesz if he would be interested in having the dot images presented and offered for sale at his gallery. Julesz agreed and told Wise about Noll's work, which was unrelated to 3-D but technically adventurous.

"I told him about my work, and he asked me if I'd like to be in an art show," Noll remembered. "He was very interested in what I was doing, but I didn't fully understand what he had in mind."

What Noll was doing, officially, was conducting research pertaining to voice communications—specifically, work on human perception of audio signal quality and techniques for identifying the pitch of spoken language. His assignments did not involve art or any aspect of the visual. Even so, the ethos of open-ended, follow-your-nose experimentation that made Bell Labs' reputation as a hothouse of technological innovation was still intact, though beginning to wane. In the early 1960s, young engineers like Noll were too new to Bell to remember the days when Claude Shannon built a mechanical mouse and ran it through a maze to see if intelligence could be made artificially, and few had worked with William Shockley when he led the team that won Shockley, John Bardeen, and Walter Brattain the Nobel Prize in Physics for developing the transistor. Shannon had departed for MIT and Shockley had gone to Stanford, where he would begin ranting vile, racist pseudo-science about the genetic superiority of the white race. After four decades of operation, Bell Labs was still a Shangri-La of Nerddom, a nurturing environment for inquisitive and ambitious engineers, especially those who were among the white males William Shockley relished glorifying.

A. Michael Noll went by his middle name, and people called him Mike. He had been educated in electrical engineering, with a PhD from Brooklyn Polytech, but had a personal interest in classical music, and he liked ballet. Narrow-framed and boyish, with a high forehead and the horn-rimmed glasses that were apparently issued along with a parking pass to all employees at Bell Labs, he looked like the third identical triplet in a family with the actor William Schallert, who played the father in *The Patty Duke Show*, and Herbert Anderson, who played the father on *Dennis the Menace.* He embodied the sweet benignity of the brainy square-guy trope of the Sixties, while quietly thinking with heterodox insight.

By 1965, when Howard Wise invited him to be in a gallery show, Noll had been experimenting for nearly three years to see if the IBM model 7090 mainframe computer at Bell Labs could be used to make images that qualified as art. The idea came by happenstance. In the summer of 1962, when Noll was a twenty-three-year-old summer intern in the audio research department at Bell, he was using a piece of equipment called a microfilm plotter, which displayed computer data as dots or lines on a small black-and-white video screen: an electronic version of the plotting device used by artists since the seventeenth century. Another researcher named Elwyn Berlekamp was taking a turn at the plotter for work of his own, and something went wrong. "I don't know what Elwyn was doing or what happened, but the plotter went haywire," Noll says, "and it started making crazy patterns with lines going every which way. Elwyn showed me, and making a joke out of the situation, he said, 'Look—computer art!'" (See photo, page 116.)

"I looked at the plotter and thought, well—hey, why not? There was a lot of talk about computer music, and Bell Labs was fairly active in that. But nobody was talking about computer art, as far as I knew. So I got to work."

Before summer's end, Noll had made enough progress to report his findings to the Bell Labs brass in a formal Memorandum for File. He submitted a portfolio of eight images he had programmed the computer to create by using pseudo-random numbers to generate zigzaggy lines in erratic designs and sets of grids with lines of uneven spacing. By twenty-first-century standards of computer-generated imagery, the renderings come across as almost comically rudimentary, not far removed from Etch A Sketch noodling. In 1962, however, the fact that they were produced by a digital computer programmed specifically for the purpose of drawing pictures made them remarkable. They had a quality of simplified modernism, too, and, without stretching things too far, they connected tangentially to the ways Klee and Mondrian foregrounded geometry in their art. (The Cubists of the early twentieth century are related to this sphere for breaking the world into geometric elements and machinelike parts, though Noll's zigzags have nothing to do with Cubism.)

The very notion of a computer making art was so new that Noll felt compelled to dodge any claim for the aesthetic value of the images, despite his aesthetic intent. "The digital computer is presently being used to produce new musical sounds and techniques of composing," Noll wrote in his memo, dated August 28, 1962, under the heading "Patterns by 7090—Case 38794–23."

> The advent of microfilm printing used in conjunction with a digital computer allows similar excursions into the field of visual art.
>
> Rather than risk an unintentional debate at this time on whether the computer-produced designs are truly art or not, the results of the machine's endeavors will simply be called "patterns."

With consent from his bosses falling short of encouragement, Noll continued experimenting with the computer and microfilm plotter, making enough images for the Wise Gallery to have seventeen to show, along with eight data-chart graphics made by Béla Julesz for his research work. Howard Wise wrote up a contract for Noll and Julesz, and they passed it along to their supervisors, who did not approve.

"There was a legal issue that was really a political issue and tied into a philosophical issue," Noll explained. "Bell Labs was half-owned by AT&T, which was a regulated monopoly in the communications business. They didn't want their people having anything to do with art. They thought art was frivolous and a waste of time. Part of this was a worry that AT&T would get in trouble for spending telephone company customers' money on something silly and wasteful like art.

"The other part of it was a weird idea my supervisors had about art. They thought something had to be worthy of being in a museum in order to be art. They were applying an aesthetic judgment on what we were doing, saying, 'You can't call this art, because it's not good enough. It's just technology. It's not creative enough to be art.' They told us, 'Stop the show—no show.'"

After some haggling over a couple of weeks, the lab bosses settled on a compromise, with terms intended to distance Noll and Julesz's images from both Bell Labs and the fine art museums the lab supervisors exalted.

The show could go on, with no mention of Bell Labs anywhere and no reference to the images as works of art. "That's why the show was titled 'Computer-Generated Pictures,'" said Noll.

"I knew the labs didn't like any association with art, and that's why I originally called the computer images 'patterns' in my technical memorandum. But I made them specifically to try to make art. Art was my purpose, my intent. Béla didn't want his data charts to be called art, because he did them purely for research. I did mine purely to make art—or, I should say, to see if a computer could make art."

To formalize Bell's detachment from the images Noll had made, his supervisors instructed him to copyright the work in his own name, which he did at first by putting a copyright symbol and his name on printouts of the patterns. He followed up by filing for copyright registrations with the Library of Congress, only to find the submission rejected. "They came back and said, 'Sorry, but this was done by a computer, not a human being. We only register copyrights done by people, not by machines.'"

Noll wrote back, explaining his process: the computer had followed a set of instructions combining elements of randomness and order. Once again, the Library of Congress declined the registration. "They said, 'We can't handle randomness. If it's random, there's no human intent.'

"I wrote back and said, 'You don't understand. I programmed the computer, so I'm the creator. The computer was just my tool, like a paintbrush.' They accepted that, and I got the copyright. But the reality is actually more complicated than I indicated. The computer was a co-creator of the artwork.

"The fact is, what the machine did is the most important part of the images," Noll added. "I was no artist. I couldn't make those images. The computer did the artistic work, using the mechanism of randomness."

In press materials for "Computer-Generated Pictures," publicists for the gallery made a point to take up the matter of authorship in this art made through the interaction of people and computers.

This exhibition demonstrates, to some small degree, the potentialities of the computer as a tool in the service of the artist. As computer technology progresses and costs come down, this technique will be more fully explored by the artist.

Presently, computer-generated pictures are limited solely by the state of the computer and microfilm art. Noll and Julesz see the day when a computer can draw—or paint—almost any kind of picture in any one or combination of colors.

Both scientists stress, however, that the artist need not fear being automated out of existence; rather, as they see it, the computer will free the artist for creation, unburdened by the tedium of the mechanics.

"Computer-Generated Pictures" opened on April 8 and ran until April 24, 1965. The Wise Gallery was a fitting environment. It had a spare modernist interior, designed by Philip Johnson, with slidable floor-to-ceiling plates covering the plaster walls and molding. As the artist Frank Gillette described the space, "It possessed an insouciant elegance that yet remained coolly unobtrusive toward the work exhibited. You entered a chamber appointed with all the elements of a public domain and simultaneously you felt like you were entering someone's very comfortable, private salon." The works by Noll and Julesz were hung in uneven, random-looking arrangements, with some too high for the average person to see comfortably, and some too low.

No attendance records have survived. No one bought any of the works, though all were available for sale. The small handful of critics who covered the show were cool to the art but not unkind. Writing in the *New York Times*, Stuart Preston seemed to struggle to come to terms with what he was seeing and defaulted to the content of the press handout, mirroring the main points of the publicity materials. "The wave of the future crashes significantly at the Howard Wise Gallery, which has on view paintings conceived by two scientists, Béla Julesz and Michael Noll, and executed by IBM #7094 computer with the assistance of General Dynamics SC-4020 Microfilm Plotter," Preston wrote.

So far the means are of greater interest than the end, this revolutionary collaboration resulting in bleak, very complex geometrical patterns excluding the smallest ingredient of manual sensibility. No matter what the future holds—and scientists predict a time when almost any kind of painting can be computer-generated—the actual touch of the artist will no longer play any part in the making of a work of art.

When that day comes the artist's role will consist of mathematically formulating, by arranging an array of points in groups, a desired pattern. From then on all will be entrusted to the *deus ex machina*. Freed from the tedium of technique and the mechanics of picture-making, the artist will simply "create."

In visual art, things had changed considerably since the days of Zoe, the automated girl, drawing portraits of celebrated Englishmen—an artificial person mimicking the way real people treated other people in art. However simple they were, the prosaic graphics in "Computer-Generated Pictures" looked like art made by a computer and an electronic plotting device. Their aesthetic was electromechanical. They betrayed no attempt to draw as fine artists are generally expected to do, but, rather, pointed to possibilities in what machines can do uniquely—if not on their own, in collaboration with human artists. As the art dealer Achim Moeller said of Howard Wise, "He foresaw the future of art to be an alliance between artistic and technological concerns, in which the machine was paramount."

"Rules for action—I was a young mathematician, so I understood that. I knew all about rules for action, but art was a mystery to me," said Frieder Nake, the German mathematician and digital innovator who made groundbreaking computer art that paralleled the experiments of A. Michael Noll in the same period of the 1960s. "The idea that I could use mathematics to program a computer to make art—that, I did not understand. Not at first."

In 1962, when Noll was beginning to test out the microfilm plotter at Bell Labs, Nake was a twenty-four-year-old graduate student at the Technical University of Stuttgart, a respected center for early digital research in the city where Nake grew up. While he was working part-time in the Recheninstitut (computer center), the school acquired a new type of equipment: a flatbed drawing machine (the Zuse Graphomat Z64) that, according to its manufacturer, could render images generated by a computer. Nake was assigned the task of developing a program to coordinate the center's equipment—to teach the computer to draw and train the flatbed device to follow the computer's orders. The challenge of breaking down visual imagery to the terms of code was acutely appealing to him, because the proposition overlapped with ideas he had been thinking about. Nake had been regularly attending lectures by Max Bense, a charismatic philosopher and intellectual provocateur, at the nearby Institute for Epistemology and History of Science, which Bense had founded. In his talks, Bense expounded on what he called information aesthetics, a set of theories that argued for understanding language and art as systems of statistically measurable units and symbols through study of the mechanics of their organization. The literal meaning of the words or images was unimportant. Through information aesthetics, then, this sentence could be analyzed for the use and recurrence of particular letters and syllable constructions in the language, with no attention paid to the authorial intent implicit in the presence of the specific words *obtuse* and *unilluminating.*

As Nake would recall in an interview conducted more than fifty years later, "Bense, in a sense, was responding to the postwar climate in Germany. He was calling for a rejection of emotion in aesthetics, which grew from a rejection of emotion more broadly. He was saying, 'Look at what happened when we Germans allowed ourselves to be manipulated by our emotions. Look at the terrible consequences. Now, be logical, be rational. Don't be emotional, even in the way you think about art.'

"Bense's purpose was heroic in its way, even though it failed miserably. He believed in the radical idea that aesthetics were wholly contained in the object and quantifiable. No subjective response, no personal response, no feeling, were permissible. Aesthetics were to be matters of measure rather than judgment, number rather than emotion, science rather than psychology. It was aesthetics perfectly adaptable to computer programming. I found it fascinating, as a computer programmer."

Nake drew useful inspiration from Bense's attention to the elements of the art object and its value as an object, irrespective of its maker's intentions or its viewers' interpretations. Feeling free to coax the computer to make whatever art it could make, Nake applied his knowledge of mathematics to employ multiple methods of generating pseudo-random numbers and multiple distribution functions—differing ways to produce materials for the art and differing ways to use them.

While Nake was experimenting, producing sets of images no one outside the Recheninstitut saw, another acolyte of Bense, a mathematician for the Siemens manufacturing company named Georg Nees, was using the computer at his workplace to make printouts of simple, attractive designs he created. Nees's work was primarily the product of his own imagination, given form through use of the computer as a tool, rather than work generated largely by the computer using its data-processing ability (the computer's "brain"). But it was produced by use of a computer, and it showed no sign of the human hand or the human heart at work. On February 5, 1965, Bense hosted an exhibition of a small selection of Nees's images in a seminar room at his institute, and began the event with a lecture titled "Principles of Generative Aesthetics," followed by a talk by Nees about his programming methods. The event took place two months before the opening of "Computer-Generated Pictures" at the Howard Wise Gallery, although no one involved in the New York show knew anything about the Stuttgart event, and no one at the Institute for Epistemology and History of Science knew what A. Michael Noll and Béla Julesz had been doing at Bell Labs.

Several professors from the Technical University of Stuttgart, along with a dozen or so students and others, listened to the presentations in evident confusion and stared at the printouts of graphical designs hung on the walls of the classroom. "No one understood a word of what Bense was saying, and I can't say I understood most of it, either," recalled Frieder Nake. "I was caught in Bense's spell, but didn't comprehend a lot of what he said. It was practically like his language was a demonstration of information aesthetics, and the meaning of the words was not important."

At the conclusion of the talks, one of the professors addressed a question to Nees: "It's nice and truly interesting, what you tell us, young man. But I wonder, can you make your machine draw like me? Can you cause it to imitate my ductus?"

Nees considered the question for a moment, then replied, "Yes, of course I can"—and, after a pause, added, "if you tell me how to do it."

With this, the professor who had asked the question walked out, accompanied by a group of others, who were soon followed by most of the rest of the people in the room. Nake watched in puzzlement. "It took a while for the cleverness of Nees's answer to sink in," Nake said. "Bense was talking about generative aesthetics, but Nees understood that computers at that time were not capable of generating art independently. They generated output in response to the input of their programming. Nees and I were working with the computers—what we made relied on computers. But, at the same time, the computers relied on us."

The exhibition space emptied out to the sound of cranky muttering. As people hustled down the hall on their way out of the building, Bense ran behind. "Please gentlemen, stay here!" he shouted, as Nake would recall. "What you see on the walls is only artificial art!"

Bense, in frustration, "coined a new phrase," Nake said, "and it marked a major change in the way people would think about art in the computer era. From this point on, there would be artificial art and natural art. Newton started this, by attempting to explain the natural world by artificial methods. Now the process was complete. Now, there could be artificial intelligence and natural intelligence.

"The problem is, the artificial is so powerful and has such a grip on us that once it's here, we care less and less about the natural or the real. This has happened. The artificial feels more real to us than anything, and the natural starts to feel unnatural. I saw this begin that day, with Bense screaming out, 'It's only artificial art!'"

Nake continued his experiments with computer-generated art, sharing the results with Bense, who was impressed by both the work and Nake's ability to discuss the mathematics underlying it. Over the following months, the curators at the Wendelin Niedlich gallery in Stuttgart, a small space recently opened, invited Bense to present a follow-up to the exhibition at his institute, and Bense welcomed Nake to join Nees for a new show billed as "Computer-Grafik." Each of the artists provided eight to ten pieces. For the art he contributed, Nake thought of Paul Klee. He wrote an algorithm to emulate aspects of Klee's aesthetic, employing pseudo-randomness to guide the program in making content choices.

"Klee became my source of inspiration as I was writing programs for art for the show, because of his maxim that the invisible should be made visible," Nake remembered. ("Art does not reproduce the visible; rather, it makes visible," Klee asserted in the prime of High Modernism.) "I was working in algorithms—abstractions impossible to see. But that doesn't mean they don't exist, and the fact that the artworks are created by computer does not make them unreal. My job was to help give expression to the ideas algorithms were capable of containing—to help the computer express itself.

"Preparing this artwork, being part of this funny little world of the first computer art, I could see that something was happening that was unusual, and I was happy and proud to be a part of it. But I thought of what I was doing, of computer art—generating art by algorithm—as something fascinating and exciting for me, but marginal. I thought it would come and go."

10

SOME MORE BEGINNINGS

I t was 1934, the age of propeller planes and vacuum tubes, when the Museum of Modern Art first celebrated the coming together of machinery and artistry by staging the Machine Age exhibition; and it was thirty-four years after that, in the time of space flight and transistors, when the same museum announced the conclusion of the era Philip Johnson had once helped it define. On November 27, 1968, a new show called "The Machine as Seen at the End of the Mechanical Age" opened at MoMA. Scheduled to run until February of the following year, the exhibition presented more than 220 objects, along with presentations of related films and other materials—all, the museum announced, in honor of "the mechanical machine, the great creator and destroyer at a difficult moment in its life when, for the first time, its reign is threatened by other tools."

Marking a distinction between the "mechanical machine" and "other tools," MoMA was alluding to the preeminence of electronic machinery in industrial societies by the late 1960s: color TVs, pocket radios, portable record players, electric typewriters, electric guitars, electric toothbrushes and

shavers . . . and, most recently, the electronic computers that were beginning to come down enough in size to start appearing along the walls of
the finance departments at major businesses. In the future, as the *Star Trek*
TV series taught us, whole planets would operate under the control of electronic "brains" so advanced as to respond to voice commands. The traditional mechanisms of gearing, tubing, and physical dynamics that had long
kept the wheels of industry and culture churning, influencing creative artists in myriad ways, were passé; indeed, they had begun to be supplanted by
electric equipment long before the MoMA show. (Movies had synced sound
forty years before 1968, and radio broadcasts and electrical recording were
flourishing at the time of the 1934 show.) The museum's timing was exemplary for a big, high-powered, deeply ingrained American cultural institution: MoMA discovered the Electronic Age only three or four decades late.

"We are surrounded by the outward manifestations of the culmina-
tion of the mechanical age," proclaimed the curator, Pontus Hultén, in the
museum's announcement of the show. "Yet, at the same time, the mechanical machine—which can most easily be defined as an imitation of our
muscles—is losing its dominating position among the tools of mankind;
while electronic and chemical devices—which imitate the processes of the
brain and the nervous system—are becoming increasingly important."

Hultén, a cherubic Swede with presciently acute taste in American art,
had been director of the museum of modern art in Stockholm, the Moderna
Museet, since 1960, when he was thirty-six. He built his reputation as a cham-
pion of young artists in later-period Abstract Expressionism and Pop Art,
introducing European museumgoers to Robert Rauschenberg and Jasper
Johns, and giving Andy Warhol his first solo museum retrospective. In 1966,
Hultén made international headlines with the exhibition titled "Hon—en
katedral" (She—A Cathedral), a fantastical indoor installation that filled
the Moderna Museet. Conceived by the feminist gadfly Niki de Saint Phalle
in collaboration with the artists Jean Tinguely and Per Olof Ultvedt, it was
constructed with steel and chicken wire, sheathed in fabric, and painted
in splashing bright colors. The structure was that of an enormous, 80-foot
long, six-ton woman—pregnant, lying prone, with her legs spread apart.
At the location of her vagina, there was a door-size opening for visitors to
enter. Inside the body, people would find a fairground of amusements and

provocations: a pond stocked with live fish, a twelve-seat movie theater, an art gallery with pretend art, vending machines, a pay phone, a playground slide for the kiddies, and beneath the construction's giant breasts, a milk bar. Hultén relished the attention the installation brought the museum, navigating the press with aplomb. Before he would start talking about machines and the muscular system for an American museum, Hultén proved deft at handling bold ideas about art and the human anatomy.

After the sensation of "Hon—en katedral," MoMA invited Hultén to curate an exhibition that could somehow follow up on the Machine Art exhibition of 1934. Taking that show as a starting point, Hultén cast the present moment as a finishing point and developed what became "The Machine as Seen at the End of the Mechanical Age." In his curation, Hultén took an independent course, making no allusion to the earlier project in his selections for exhibit content or the accompanying text. For the printed catalog, Hultén wrote 216 pages of historical essays and descriptive prose on the items he curated, which he called a "historical collection of comments on technology by artists of the Western world." He made no mention of Philip Johnson, who had by now achieved the world fame he had long strived for as an architect of icy hypermodern buildings. The book was bound with front and back covers made of embossed stainless steel, fastened with metal hinges.

Hultén brought historical sweep to the catalog text, touching on the philosophies of both Descartes and La Mettrie, along with Philo of Byzantium, Hero of Alexandria, and the Roman architect Vitruvius—but without tying in Plato, as Johnson and his co-curator, Alfred H. Barr Jr. had done. (Hultén worked in a bit about Aristotle.) Every item exhibited in the show was shown in a photograph accompanied by prose by Hultén that was often more analytical than descriptive, conveying not only how the machine was seen at the time the work was made, but what Hultén saw in the occasion of that seeing. For example, the exhibition featured a specimen of Jaquet-Droz's automata, the robotic writer. (The catalog attributed the work solely to the elder Pierre Jaquet-Droz, rather than to the father–son team of Pierre and Henri-Louis Jaquet-Droz.) "To contemporary spectators, the little mechanical writer must have seemed almost intolerably perfect," Hultén wrote. "He must have inspired feelings of curiosity, admiration, and probably also paralyzing inferiority. The young scholar embodies the idea of perfection—an ideal man, who never makes an error, never gets in a bad humor, and never revolts."

Apparently taken by the phenomenon of automata, Hultén brought the subject up in promotion for the exhibition as well. "They posed a riddle," Hultén said, as he was quoted by the critic Barbara Gold in the *Baltimore Sun.* "What was the distinction between man and the inanimate beings that moved and functioned like man? There was something intriguing in the sacrilegious idea that these were men created not by God but by man himself, and thus without souls."

Throughout the catalog text, Hultén provided sharp analyses of the works he had selected, mingled with ruminations and digressions related to machines, art, and other interests of his. In a passage about a satirical mechanical sculpture by Jean Tinguely, Hultén delivered a short lecture-essay on the dangers in the mechanization of art history and the art market:

> The industrialization of the New York art scene, where the journalists who write about culture expect movements in art to succeed one another like each season's fashions, is a phenomenon we have witnessed in recent years. Such mechanization of art history is all the more unnecessary, because it is totally unproductive and arises out of laziness and lack of imagination. Free expressions that have been created through individuals, reactions and inventiveness are forced into an artificial pattern of development, because it is easier to present them in that way. This mechanization is also part of the rapidly accelerated commercialization of art that has occurred in New York and elsewhere in the past few years. Art is viewed as a consumer product and sold on that basis; therefore, at the opening of every season a new model must be unveiled.

The show was promoted as a historical overview of commentary on machinery and technology by artists, and it was populated mainly by works conceived as art, as opposed to the utilitarian devices with aesthetic quality that had distinguished the 1934 show and the *Little Review* exhibition preceding it in 1927. Commentary in the forms of both approbation and critique was clearly the purpose of a great many pieces in the 1968 show, from the ecstatic renderings of speeding automobiles by Italian Futurist painters and the found-material sculpture of Dadaist Raoul Hausmann to the wild cartoons of impossible goofiness by Rube Goldberg and more. A lovely painting by the British commercial illustrator W. Read brought a Turner-like sense of heavenly awe to a rendering of the interior of a gaslight factory in the mid-eighteenth century: technology as a source of literal and spiritual illumination. A dark, bleak, angry painting from 1942 by the American artist Mark Tobey suggested a world despoiled by industrial and household machinery. Its title: *The Void Devouring the Gadget Era*. A monstrously elaborate mechanical structure by Tinguely, simultaneously whimsical and menacing, spewed out toy balls that museumgoers could catch and throw back into the construction; the act, to Tinguely, was a way to ridicule the perpetual cycle of mass-producing useless goods.

Unlike the 1934 show and its predecessor, the 1968 exhibition presented only a handful of actual machines, machine parts, or photos or schematic diagrams of them. The passive-voice "Machine as Seen" design of the project centered on what artists saw in machines, rather than the machines the larger population could see all around them. There was a photo of a massive steam-powered locomotive from 1771, notable for historical interest as "the oldest self-propelled vehicle in the world." More noteworthy for aesthetic value were two full-size working vehicles: an extravagantly gorgeous Bugatti Royale from 1931, designed in grand, sweeping curves, and a cartoonish three-wheeled Dymaxion car designed by Buckminster Fuller and built in 1933. The particular Dymaxion model on display was a duplicate of the original, which had been destroyed in a road accident that killed the driver.

The meatiest part of this exhibition was the substantial collection of artworks that served to demonstrate the profound influence of machines on visual art in the modern era. Well over a hundred paintings, drawings, sculptures, and other works in the show made cases of varying kinds for the impact of machinery, technology, processes of industry, and industrial culture more broadly on the way artists had seen the world and humankind's place in it. "The Machine as Seen" was the predicating proposition of the exhibition, but it showed how differently artists came to see everything else because of machinery.

The show offered ample evidence of the machine's influence on historical art movements, demonstrating how Cubism converted natural forms into the geometric designs and mechanical-drawing angles of industrial art; how photography, camera lenses, and lighting for both still photos and motion pictures changed the way artists in other visual fields framed and constructed imagery, especially in the era of Expressionism; how the artifacts of reproductive processes and commercial printing could be employed in the collage form; and how machine parts and even whole machines were incorporated into artworks or repurposed as art.

Hultén made persuasive arguments for the importance of machinery in the works he curated. In the descriptive text for *The Crowd*, a painting from 1915 by the English artist Wyndham Lewis, he wrote, "Forces of construction and destruction are juxtaposed and in battle. The painting is built up of zones of action, defined by different kinds of mechanical and geometrical forms, new and unseen in architecture at that time. Instead of human beings, there are a few robot-like figures."

Similarly, in a catalog passage on Surrealism, Hultén teased how "the estrangement from mechanical things and the fear of machines that were later to predominate the art of the Surrealists. The core of the Surrealist program was exploration of the inner depths of man's mind, and for that, there was no need for machines—at least, for any known kind of machine. To the Surrealists, the world of technology represented an intrusion, if not an actual menace."

As one writer suggested in a review of the show, machines probably informed art-making in even more ways than Hultén addressed. The critic Christopher Andreae, writing in the *Christian Science Monitor*, pointed out that methods of mechanical production were no longer uncommon in art studios, and he was not talking about artists experimenting in new technological branches of art such as kinetic art, sound art, and computer art. "A thing which is perhaps not emphasized by this exhibition is the fact that many artists today, not at all concerned with machine imagery, nevertheless look to technology for tools in making their works," Andreae wrote. "They use power tools—commonplace ones perhaps, but it is a significant point. A number of painters now use spray mechanisms; some sculptors have their sculpture factory-built for them; interest is being shown in the possibilities of mass production as a way of escaping the persistent belief in the sanctity of the unique art of object. The effect of mechanization has not been confined to those artists who have been concerned to 'comment on technology.'"

No role in art-making was necessary at all for machines to have a place in the realm of art, argued the critic John Canaday in the *New York Times*. "Machines are beautiful as themselves, independently beautiful," Canaday wrote, "and I suspect that the best shows, the most beautiful shows, related to the kind of thing now on view in Brooklyn, take place in no museum but in laboratories—just as, in the Museum of Modern Art's show, none of the paintings or sculpturers of the last half-century can rival the beauty of the machines that inspired them, as a revolt against technology or an effort to fuse with it."

Though Canaday found much worth seeing and thinking about in the MoMA show, he was more impressed by a related exhibition at another museum. "It takes a bit of doing for the Brooklyn Museum to make the Museum of Modern Art resemble an archeological dig, but that's what's happened," Canaday wrote.

The two museums had been working in cooperation to present experimental new art from the same source, an unusual collaboration by institutions generally thought to be rivalrous. While MoMA included this work as one component in a historical overview, the Brooklyn Museum gave it a showcase of its own, and while MoMA oriented its presentation on "The End of the Mechanical Age," the Brooklyn Museum called its show "Some More Beginnings." The common source of content was a recent art competition sponsored by the not-for-profit consortium Experiments in Art and Technology or, *très*-Sixties style, E.A.T. Organized the previous year, E.A.T. was a joint venture by two engineers from Bell Labs, Johan Wilhelm Klüver (who went by the nickname Billy) and Frederick Waldhauer, and two artists, Robert Rauschenberg and Robert Whitman. Their charter, drafted by Klüver and Rauschenberg, established E.A.T. with four lofty and vague goals:

Maintain a constructive climate for the recognition of the new technology and the arts by a civilized collaboration between groups unrealistically developing in isolation. Eliminate the separation of the individual from technological change and expand and enrich technology to give the individual variety, pleasure and avenues for exploration and involvement in contemporary life. Encourage industrial initiative in generating original forethought, instead of a compromise in aftermath, and precipitate a mutual agreement in order to avoid the waste of a cultural revolution.

The competition called for teams of parties, one an artist in any discipline and one an engineer in an emerging technology, to try to produce "the most inventive use of new technology." Which technology (or technologies) in particular to employ was up to the contestants, and so was the form their "use" would take. By an arrangement with the two museums skewed to the one in Manhattan, the work by winners of the first prize award ($3,000) and the two second prizes ($1,000 per team) would be shown as part of the Museum of Modern Art's "End of the Mechanical Age" show, along with six more works from the competition selected by MoMA in advance of the judging. From the pool of 147 entries submitted, a selection of 137 would become the contents of the Brooklyn Museum's "Some More Beginnings" exhibition. (For reasons unclear, the terms of the awards specified that they would be granted to the engineers alone, although all the collaborators were credited in the museum displays and catalogs.)

The first-prize work, shown at MoMA, was an interactive sculpture by Jean Dupuy, a French-born artist, and Ralph Martel, an American who was listed as the participating engineer but was a fine artist, educated at Cooper Union, who had some technical savvy. Titled *Heart Beats Dust*, the piece read sonic input provided by onlookers—their heartbeats, their speech, the sounds of their feet—and transferred it onto clear plastic that produced static electricity and made patterns out of dust. As Hultén noted, "Like many works of recent years, *Heart Beats Dust* manifests a new form of cooperation with nature . . . a sensitive collaboration between natural forces within and outside the body."

Other pieces submitted for the E.A.T. competition also worked interactively, with sensors picking up input from museumgoers and using it to trigger an artful display of some kind. An outsized structure by artist David Lawner and engineer Garry Furgeson took sounds made by people observing the piece and used them to make blinking patterns in an array of lightbulbs. With a work called *Fakir in 3/4 Time*, the artist Lucy Jackson Young and the engineer Niels O. Young simulated a variation on the Indian rope trick, using a motor and a fan to make a long loop of fabric tape twirl in the air. With *Proxima Centauri*, the artist Lillian Schwartz and the engineer Per Biorn offered an exceedingly complicated and eerily striking kinetic sculpture—Schwartz's conception, with aid from Biorn in executing the necessary equipment. Outwardly, the piece appeared at first to be no more than a rectangular black box, 55 inches high and 30 inches wide, with a plastic dome protruding from a hole on the top. It looked like a giant-size container of roll-on deodorant. As observers approached the piece to get a closer look, they would step onto a pressure-sensitive floor pad, sending signals to trigger a mechanism inside the box. The dome would slowly rise and reveal its true form as a plastic ball, while glowing abstract designs in rainbow colors purled and rippled on the inner surface of the dome. (See photo, page 132.)

The effect was transfixing, according to one of the participants in the E.A.T. competition, the Bell Labs engineer Kenneth Knowlton, whose own submission was a 12-foot print of a mosaic portrait of a reclining nude woman, the dancer Deborah Hay, which he and his fellow Bell scientist Leon Harmon made by using a computer to convert the gray tones to telephone circuit diagrams. "Mechanically, *Proxima Centauri* was a hodgepodge of things you'd find in your closets," Knowlton said. "But they were put to very creative use, and the effect was rather inexplicably hypnotic." The technical components whirring and spinning in the box included a 35-mm slide projector, a motor with a cam that stirred up ripples in a tank of fluid, a mirror, a red lightbulb, and a blue lightbulb, along with the electrical equipment Biorn helped Schwartz devise.

"Lillian was a fireball," recalled Knowlton, who first met Schwartz through E.A.T. and got to see her work after his project collaborator, Leon Harmon, invited her to join Bell Labs as an unpaid "resident visitor." She would go on to experiment imaginatively at the labs for decades, doing innovative work in computer graphics and hybrid forms of technological art. "She wasn't a person who would walk into a room and say, 'Oh, there's a computer, and there's a spectroradiometer, and there's the exit light above the door.' She looked around and saw *materials*. Whatever inspired her was valid in her eyes—mechanical, electronic, or flesh and blood, if it inspired her and could help her in her work. She was ahead of her time."

Three hundred sixty years ahead of her time, to be precise. On stardate 5431.5, by the twenty-third-century calendar used on the *Star Trek* series in the Sixties, the starship *Enterprise* will be commandeered by a miniskirted alien woman who immobilizes the crew, surgically removes Mr. Spock's brain, and takes off with it to space-parts unknown. Adventure and unintentional self-parody ensue, as Captain Kirk and his pal Dr. McCoy search the cosmos for their friend and crewmate's brain, with Spock himself in tow, walking robotically by way of a skull attachment McCoy worked up and directs with a remote control. When they determine the location of Spock's brain, in an underground command center on the planet Sigma Draconis VI—where else?—what they find is an object almost identical to *Proxima Centauri*, the kinetic sculpture by Lillian Schwartz and Per Biorn. The globe on the artwork glows eerily, to be read here as the force of Spock's extraordinary Vulcan (or half-Vulcan, half-human) mind. Six plastic rods jut out from the sides of the black box holding the globe—shooting instructions from Spock to the electromechanical and hydraulic infrastructure of the planet.

In its cheesy way, the "Spock's Brain" episode of *Star Trek* touched on a couple of ideas about the mind, the body, and computers worth thinking about. One, given form in the sight of a brainless Spock walking around, is the centuries-old concept of human (or half-human) duality of body and mind. Staring blankly and moving mechanically like Motogirl, Spock was not fully alive without his mind and body united. The newer idea, dramatized by the premise of Spock's neural activity controlling the functions of a planet from within the *Proxima Centauri* sculpture, was a radical proposition in 1968 and is still resonant in the twenty-first century: that the brain of a living person and a technologically advanced computer do the same things and are utterly interchangeable. The computer system breaks? Replace it with the best living brain you can find.

The theoretical resonance of "Spock's Brain" was dampened markedly by the overall crappiness of what is widely seen as the worst episode ever in the history of the *Star Trek* franchise. "Frankly, during the entire shooting of that episode, I was embarrassed," wrote the actor Leonard Nimoy in his memoir, *I Am Spock*. La Mettrie would have understood that brainlessness does not have to be mindless.

11

IT'S LIKE A ROBOT

Our conceptions of artifice and authenticity have always been mutually dependent. Ever since machines began to change the world and the way humans have thought about it, establishing the idea of the natural in opposition to the mechanical, new mechanized or electronic ways of doing work and making art have redefined the methods that preceded them or, in some cases, defined the old ways for the first time. Music played by an automated man in a Piccadilly music hall got listeners thinking about what it was that they had relished in music made by a living player. From then on, if real people were to make music like that of a machine, it would be thought of as unreal, untrue—not merely new or different, but, by association with machinery, inauthentic: lesser. Artworks produced by industrial methods such as the silkscreen portraits of Andy Warhol and the plywood constructions of Donald Judd, or the abstract ink drawings made by a computer programmed by Michael Noll, crystallized what art lovers had valued in the work of human hands.

Artifice alters reality—or our understanding of it. And sometimes, synthetic things become the new reality—or part of it. With time, some kinds of art and music that seemed unnatural or unreal at first become familiar and eventually commonplace. They seep into the fabric of the world we know and settle into it, become a piece of it. In music, this happened over the course of the late twentieth century and early twenty-first century, when a family of electronic technologies moved steadily from the fringes of novelty attraction and the avant-garde to the heart of mainstream culture, changing the way a great deal of music was made and what listeners made of it. The main character in this story is a device named for its ability to fabricate music artificially, producing sounds that are, by definition, synthetic.

In the fall of 1968, the term "synthesizer" was still largely unknown outside of laboratories for the electronics and chemical industries. An early Moog synthesizer was set up to serve as a freaky-looking prop for a scene with Mick Jagger in the hippie freak-out movie *Performance*, and one of the technicians on the set called it a "sanitizer," having no idea what it was. The Moog in the film was Jagger's own property, though he never used it with the Rolling Stones. The bits of synthesizer music blipping and whizzing in the background of a few scenes in *Performance* were performed by Bernie Krause, a former student of electronic music at Mills College. Krause was working in a two-hatted role as performer/sales representative for the R. A. Moog Company, manufacturer of the eye-catching and ear-stretching sound-synthesizing equipment devised only five years earlier by a thirty-year-old Cornell PhD student with the name of a wizard from Mars, Robert Moog. Krause and another musician, Paul Beaver, would tote the stove-size early synthesizers to potential users and demonstrate them and give lessons in using them, in hopes of making deals to sell them. Beaver and Krause had the wisdom to pitch creatively ambitious and very wealthy rock stars like Mick Jagger on the virtues of a bizarre new machine that cost almost $15,000 and was oppressively difficult to play but looked super-cool.

While MoMA and the Brooklyn Museum were setting up their simultaneous shows of mechanical, electronic, and hybrid visual art in New York, George Harrison was in Los Angeles, stepping away from the Beatles temporarily to produce an album for Jackie Lomax, a blue-eyed-soul singer from Liverpool he had signed to Apple Records, the label the Beatles had recently formed. Harrison, acting for the first time as producer for another artist, booked the *crème* of California studio musicians, a loose assemblage of rock pros associated with Phil Spector sometimes called the Wrecking Crew. On the last day of sessions, November 11, Bernie Krause and Paul Beaver came to the studio to show Harrison how the Moog synthesizer could be used to add unusual electronic textures and filigree to the recordings. When the sessions were finished, late that night, Harrison asked Krause to stay and give him a private demonstration of the Moog: *How many different sounds could it make? What did the operator have to do to make them?* Krause proceeded to walk Harrison through the range of electronic tones the Moog could produce—one at a time, since the device could generate only single notes, unable to play chords or even two notes simultaneously. Press one of the keys on a keyboard attached to the array of interconnected components covered with knobs, plugs, and switches, and the Moog would produce a low, rumbling growl. Turn a few knobs, and out comes a fizzy hiss. Flick a couple of switches, and the hiss would start to wiggle. Turn another knob, and the wiggle would slow down and stretch out to be a wobbly warble. Run the tip of your finger along a stretch of metal ribbon, and you would hear a siren rising higher and higher and higher in pitch. Plug in a couple of plugs, and hear a staticky, ratchety sound unlike any sound in nature. Krause kept turning, flicking, plugging, and unplugging, and out came a cartoon circus parade of random gurgles, squiggles, bleeps, chirps, honks, howls, and noises without English words to describe them properly.

Harrison was impressed and drawn to the exotic, otherworldly aura of this baffling machine for making extra-human music and sound. Famous for bringing dimensions of Eastern mysticism and Indian classical music into the sphere of the world's most popular band, he had recently produced a film score that foregrounded South Asian musicians in an aural mosaic with Western pop elements. The soundtrack album, *Wonderwall Music*, had just been released as the first solo album by any of the Beatles. Open to avenues of creativity as yet unexplored by other pop artists—including John Lennon, Paul McCartney, and Ringo Starr—Harrison placed an order to buy the same model of synthesizer he had seen and heard in Los Angeles, the Moog IIIp. He asked Bernie Krause to have the equipment shipped to the English county of Surrey, where he was living in a low-roofed bungalow with exterior walls painted in swirling patterns of psychedelic colors: wonder walls! Now, with a Moog, the inside of the house could sound like the outside looked.

"It was enormous, with hundreds of jack plugs and two keyboards," George Harrison said of his new acquisition. "But it was one thing having one, and another trying to make it work. There wasn't an instruction manual, and even if there had been, it would probably have been a couple of thousand pages long. I don't think even Mr. Moog knew how to get music out of it."

Unable to do much with the Moog on his own, Harrison summoned Krause to England for a personal tutorial. As Krause prepared to instruct Harrison in the use of frequency oscillators, control voltage attenuators, wave multipliers, and trigger sequencers, Harrison broke the news that he was planning to release a full LP of synthesizer music on Apple Records' subsidiary label for art music and spoken-language recordings, Zapple, which John Lennon and Yoko Ono were using for their latest experiments in tape loops and found sound. In fact, Harrison continued, he already had a full side of the album completed. "I want to play something for you that I did on a synthesizer," Harrison announced, as Krause would recall. "Apple will release it in the next few months. It's my first electronic piece done with a little help from my cats."

Harrison played the tape, and, after a minute, Krause recognized what he heard. Back in Los Angeles, when Krause was walking through a raw demonstration of Moog sounds after the Jackie Lomax sessions, Harrison had the studio tape recorder running without Krause's knowledge. Now, he was planning to release an edit of the demo as an electronic composition. "George, this is my music. . . . Why is it on this tape, and why are you representing it as yours?" Krause asked.

"Don't worry," Harrison replied. "I've edited it, and if it sells, I'll send you a couple of quid."

The album *Electronic Sound* was released by Zapple Records under George Harrison's name in May 1969. The front cover was a fun, brightly colored acrylic painting by Harrison himself, done in the same ebulliently naive style as the cover painting Bob Dylan had done for the debut album by The Band, *Music from Big Pink*, the previous year. Harrison's artwork depicted, in its way, a man with green skin and bright red lips, wearing a yellow bow tie, surrounded by Moog components. (In liner notes for the reissue on CD, Harrison's son Dhani said the green-faced person was his father's image of Bernie Krause.) The man is positioned behind a big control panel, the front of which is overladen with abstract designs: pink squiggles, green and white dots, something like a shooting star or a fish, and a red Jewish star—presumably an allusion to Krause—with the spout of a meat grinder on one side, spewing out sounds.

The record had two selections, each filling a side. One was the recording of Krause's Moog demonstration in Los Angeles, edited to about twenty-five minutes in order to fit on a vinyl LP and given the title "No Time or Space." That was a phrase Harrison had used in interviews to convey the experience of Transcendental Meditation. The other track had actually been generated by Harrison on his Moog: a sequence of elementary synthesizer effects, seemingly disconnected, which Harrison titled "Under the Mersey Wall." (That was a twist on the title of a column in the *Liverpool Echo* newspaper, "Over the Mersey Wall," written by an unrelated writer also named George Harrison, as well as a vague claim to underground status.) On pressings for the U.S., the titles on the record labels were inadvertently reversed. A credit line for "No Time or Space" read "Recorded with the assistance of Bernie Krause."

As Krause would later say, "I had no control over any of it. I didn't know it was being recorded. I didn't want it out, and I felt very badly that he had to do that."

Electronic Sound did not sell copies sufficient in number for Harrison to send Krause any amount of quid. Within weeks of the album's release, the Beatles' new business manager, Allen Klein—a hard-nosed philistine with the interpersonal style of a scrap-metal shredder—shut down support and funding for Zapple Records with no public announcement. *Electronic Sound* went unpromoted and virtually unnoticed, never appearing on the UK sales charts and poking up for only two weeks near the bottom of the US *Billboard* chart, peaking at number 191 out of 200.

In the *Rolling Stone* record review section, critic Ed Ward took up *Electronic Sound* briefly at the end of a piece focused mainly on the only other album released on the very short-lived Zapple label, John Lennon and Yoko Ono's *Unfinished Music No. 2: Life with the Lions.* Ward tore into

Lennon and Ono's album, the second of their three releases of sonic experimentation, which included a recording of the fetal heartbeat of the couple's unborn baby, taped on cassette before Yoko miscarried, along with an
edit of a live performance Lennon and Ono gave with a couple of free-jazz
musicians at Cambridge University. Ward dismissed it as "utter bullshit."

He was gentler with *Electronic Sound*. Unaware that another person
was playing on one of the two tracks, he gave Harrison some credit for
doing "quite well learning his way around his new Moog Synthesizer in
such a short time," adding, "but he's still got a way to go." The album "is
fair enough as a beginning effort, and although one can imagine George
sending tapes of it to his friends, it is hardly musical product," Ward wrote.
"The textures presented are rather mundane, there is no use of dynamics
for effect, and the works don't show any cohesiveness to speak of. However,
if he's this good now, with diligent experimentation, he ought to be up
there with the best in short order." In a poke at the albums' pretentiousness, the review was credited to Edmund O. Ward.

An edgier semi-underground magazine called *Fusion*, published out
of Boston, took a properly unconventional approach to *Electronic Sound*
with a bitingly mimetic review by Loyd Grossman, a Massachusetts-born
writer who would later relocate to London and become well known as a
cultural broadcaster. This was his review, in its entirety:

THIS ALBUM . . . phlurp phlurp phlurp . . . is on Zapple (grau-
ughh! *'&'%%!) the Beatles personal label /--perhaps no other-
company   wapwapwapwapwapwap   uuuhwweeoques—would
record it-+++++++++++

The ALBUM was composed (??????????) and produced (!*
@3#+"_#) by George kpowie kpowie uuuuuuuummmm
HarrisOn;,V2#$. It has zigazigazigaziga z i g a z i g a z i g a aaa two
sides and is made of AS cEEERP!EEERP!EEERPPPPPPP! and
comes with a lble(oopsO) label and a cover @@@@twerptwerpt-
werptwerp @@@ and is 4three minutes and 50+1 seconds long,
which bzzzzzzbzzzzzzbzzzawwww is a lot of ;5;5;5;5; nois-e ¢¢;#;¢.
As an extra twyehdgeysgetdfefcbruhebdtefsf bonus it is ROUND
and ooooooooooopppawwwiee water resistant S O THAT when you

are t h r u h--------- listening to it;;;it can be used as a meeooom porthole cover.. . . . . If YOU fweemfweemfweemfweem apapapapapapap have an olD Sunbeam Toaster ugwachattttatta-churgchurg churg and enjoy putting your dddddlddddlddlllder ear up to it wwhoooooooggggg*-*- you may enjoy this album phw-erpphwerp phwerp phwerp phwerp phwerp phwerp phwerp phw-erp.%.%.@V4:!*.—L@yd gr¢ssM&n

Music synthesizers have been around as long as musical instruments. In a sense that is not just playing with terminology, all instruments have always been synthesizers. As a rule, they aim to replicate sounds from the natural world—to synthesize those sounds, as a piccolo conjures the sound of birdsong or a saxophone evokes the sound of the human voice; or they aim to produce sounds not to be found in nature—artificial sounds, synthetic sounds.

Beyond the domain of such traditional instruments, meanwhile, a whole class of electronic keyboard innovations emerged in the mid-twentieth century for the purpose of emulating the sounds of as many other musical instruments as possible while, on top of that, generating new kinds of sounds no other implements could produce. They had antecedents in the dozens of makes and models of opulent music machines popular in the nineteenth century and early twentieth century: the orchestra boxes such as Rex, the Popper company's Orchestrion that's still entertaining school trip audiences at the Morris Museum in New Jersey. In time, the electronics of the vacuum-tube era led the art of music synthesizing to advance considerably, and wonders of the Mechanical Age like Rex took on a new kind of wonderment as antique curios.

The first electronic keyboard instruments designed to synthesize the sounds of multiple other instruments were essentially smaller, vacuum-tube versions of the mammoth, intricately engineered pipe organs that had long brought the variety of aural colors, the volume and force, and the grandeur of the symphony orchestra to cathedrals and other houses of worship, concert halls, and one major department store, the block-size Wanamaker store in Philadelphia. (The role of popular retail outlets as commercial parallels to houses of worship in America has been obvious since the heyday of department stores in the mid-twentieth century.) The Wanamaker organ, still working after 120 years in operation and played regularly in the third decade of the twenty-first century, is said to be the largest fully functional pipe organ in the world. Built for the 1904 World's Fair in St. Louis and later purchased by John Wanamaker to draw floor traffic in his store, the organ is monumental in scale and complexity, with six keyboards and more than a hundred hand and foot controls, 27,750 pipes in 464 configurations spread out in space in five floors of what was once the Wanamaker building, now owned by Macy's.

In a public talk before a tour of the organ and performance in January 2020, a representative from the store recited a litany of impressive factoids about the instrument and its history. "Leopold Stokowski was so inspired by this organ that he commissioned works to be played on it," the guide said. "Composers of note like Richard Purvis and Marcel Dupré were so moved by it as to compose works specifically for the organ. A number of notable music works wouldn't exist were it not for this organ. The organ didn't create the music, per se, but it inspired the compositions.

"It takes several months of dedicated study of this organ and practice to be prepared to play it to its full capacity," he added a few minutes later. "There are currently only two or three musicians in the country qualified to sit down right now and play it to capacity. It's not like playing a Casio." That a small Casio keyboard could simulate a great many of the Wanamaker organ effects is an additional fact the guide did not mention, though no electronic instrument of the digital era could be expected to come anywhere close to competing with the resonance and dynamic impact of 27,750 pipes spaced over five floors of a building.

The road from the Wanamaker organ to Casio keyboards is dotted with technological landmarks, and the Moog synthesizer is surely foremost among them. More than twenty years before the Moog, however, another electronic synthesizer, now all but forgotten, received a good amount of attention and deserved it. Invented by Laurens Hammond, who also developed an electric organ and marketed it through a company carrying his surname, the first popular synthesizing instrument was called a Novachord, a fitting name for a machine that produced electric-textured polyphonic chords designed to suggest the eruption of a nova.

(Once again, the first Moog synthesizers would not be capable of playing chords at all.)

The term "synthesizer" was not in the musical vocabulary when the Novachord was introduced, in 1939, but the concept of synthesis was in the air. The synthetic products of synthetic processes had entered the national conversation, with human-made fibers beginning to transform the garment industry needed for parachutes. A synthetic silk called nylon was appearing in women's stockings just as wartime rationing was limiting supplies of natural silkworm products. Separately, a pair of British chemists, James Tennant Dickson and John Rex Whinfield, developed a new polyester material for use in fabrics. The American company Dupont bought the rights to it, twiddled with the formula, and introduced it under the brand name Dacron in 1941. In the same period, advances in pharmaceutical chemistry were leading to the development of groundbreaking synthetic drugs: synthetic hormones and antibiotics. Chemists and clothing manufacturers and medical scientists were all busy synthesizing. Why not musicians?

The Hammond company introduced the Novachord to the public in a futuristic setting: the Ford Pavilion of the 1939 World's Fair, with the composer Ferde Grofé conducting four musicians playing three Novachords and a Hammond organ. Early observers found the instrument "amazing and startling," the *New York Herald Tribune* reported, explaining how it "resembled the piano but can be played to stimulate the sound of nearly any musical instrument," from "the brassy blare of the trumpet" to "the clear call of the English horn" and other less alliterative phenomena. As the composer John Cage would note, "Most inventors of electrical musical instruments have attempted to imitate eighteenth- and nineteenth-century instruments, just as early automobile designers copied the carriage."

Laurens Hammond stressed the more distinctive ability of the Novachord to produce unique, unconventional electronic sounds. As an Associated Press report pointed out, "Its inventor does not consider it an imitative instrument in the sense that it would be substituted generally for any of the instruments whose voices it can produce. His idea is that it brings to music entirely new possibilities which will be developed for varied orchestra effects and for greater diversification and interest in home entertainment."

This approach made good sense politically as well as aesthetically. Almost immediately after news of the Novachord broke, the existence of a single instrument capable of doing the work of a dozen others triggered panic at the American Federation of Musicians. The union announced a sweeping prohibition of public use of the Novachord by its members. Within weeks, Hammond lobbied successfully for a compromise whereby an AFM musician would be permitted to use the instrument professionally under a set of detailed conditions, such as only if the Novachord were not taking the place of a different instrument and only if the Novachord were not being used on a type of job that had previously called for the use of multiple instruments. The fear of machines taking over for humans is one of the great constants in the history of technology, and it is equally easy to inflate or dismiss.

In part because it didn't really sound very much like a trumpet, an English horn, a Hawaiian guitar, or any of the other instruments it sought to emulate, the Novachord was generally used for the wholly unique sounds it could produce. The British singer Vera Lynn, heartthrob of the Kingdom's boys in uniform, had a major hit with a ballad of romantic yearning, "We'll Meet Again," sung to the accompaniment of a Novachord providing a wash of supernatural-sounding tones. With the spectral quality the Novachord brought to the recording, Lynn seemed to be assuring both soldiers and loved ones waiting on the home front to take heart: they will be reunited one day, if only in the afterlife. (On the record label, the singer, the keyboard player, and the instrument received roughly equal billing, with the names of all three printed in capital letters the same type size: VERA LYNN with ARTHUR YOUNG ON THE NOVACHORD.) Along the same lines, the film composer Franz Waxman used the Novachord for ghostly effect in parts of his score for Alfred Hitchcock's *Rebecca* in 1939, and Max Steiner employed it in a similar manner, though more subtly, in the *entr'acte* of his music for *Gone with the Wind* in the same year.

Pop and jazz composers with advanced senses of humor used the startling tonalities of the Novachord in clever ways, leaning into the instrument's oddity. Slim Gaillard, a composer/singer/multi-instrumentalist known for blending improbable combinations of ingredients on recordings like "Mishugana Mambo" and "Sukiyaki Cha Cha," showed that the strange new machine could swing with a juke-joint kick on his record "Novachord Boogie." In a kindred spirit, a virtuoso whistler and rhythm-bones player named Freeman Davis, working as Brother Bones, revived a standard of the Charleston era, "Sweet Georgia Brown," with Nova-chord backing. The record was an unpredictable hit of unlikely durabil-ity, played countless times for decades as the theme song of the basketball comedy act the Harlem Globetrotters.

The Novachord, made with 163 vacuum tubes and more than a thousand capacitors hand-wired and individually soldered, was time-consuming and expensive for Hammond to produce, and costly for users to purchase and maintain. Encased in a cabinet of cherrywood, each instru-ment weighed about 500 pounds and cost $1,900 in 1939 (about $24,000 today). The price of a Novachord was almost exactly half the average cost of a new home in 1939. After three years on the market, only 1,069 Nova-chords were sold, and manufacturing expenses were rising with wartime restrictions on electronic components. Hammond discontinued the Nova-chord in July 1942, shifting its focus to the electric organ market.

After the war, the Hammond company worked at developing an electric organ that would carry its own identity as an electronic instru-ment. That meant, in part, emulating a range of other instruments—synthesizing a variety of sounds—as pipe organs had always done, but with qualities unique to an electronic machine. In earlier years, Ham-mond had been forced to confront the folly in attempting to mirror the performance of pipe organs too closely. The company had been advertising its electric organs as capable of producing "real" music that was "fine" and "beautiful," producing "infinite tones" capturing "the entire range of tone coloring of a pipe organ." The Federal Trade Commission, responding to complaints from the Pipe Organ Manufacturers Association, filed a charge of false and misleading advertising against the Hammond organization.

In protracted hearings on the case, the FTC adjudicated a comparison test of a $2,600 Hammond Model A electric organ and a $75,000 instru-ment with 6,610 pipes, built by an esteemed organ maker, E. M. Skinner, for the Rockefeller Chapel at the University of Chicago. Two juries, one of music experts and one of student volunteers, assembled in the chapel pews to hear thirty selections of music and name the instrument they thought was playing each one. The test, though not rigorously scientific, suggested that distinguishing between the two organs might not be easy for every listener. In response, the pipe organ trade group claimed that Hammond's people had tweaked the settings of the Skinner instrument to make it sound more like a Hammond. The FTC, in its ruling, made a compromise. It called for Hammond to cease making claims of parity between its electric products and pipe organs, but did not prohibit the company from continuing to call its music "real," "fine," and "beautiful." In a rare incursion by the commission into the domain of aesthetic values and the ideology of authenticity, the FTC allowed that the output of electromechanics could be something real, and that the sounds could be fine and beautiful in ways of their own.

The Hammond company recovered well from the collapse of its Novachord adventure, refining its line of electric organs and finding them selling steadily. Its best customers were churches, particularly houses of worship in African American communities without the space to house or the means to purchase big, expensive pipe organs. The vibrantly electrical sound of the Hammond, with settings that brought a piercing bite to the chordal attack, clicked with musicians and congregations in Black churches. By the mid-1950s, hundreds of Hammond B3 organs were selling to churches every week. The B3 became the sound of gospel music in the critical postwar period of the civil rights movement, when the Baptist church was a key mobilizing force for Black consciousness and activism. Dr. Martin Luther King Jr. preached with an electric organ in the background at Dexter Avenue Baptist Church in Montgomery, Alabama, and the Rev. C. L. Franklin led his followers to gospel music played on a Hammond at New Bethel Baptist Church in Detroit. Franklin's daughter Aretha would sometimes play that organ on Sundays, and she would grow to take the fire and open emotionality of gospel music far beyond the church walls.

An impassioned new style of popular music grew directly out of Black gospel, as Aretha Franklin, Sam Cooke, Ray Charles, and their peers adapted works of sacred art to secular purpose, transmuting "This Little Light of Mine" into "This Little Girl of Mine," and "Talkin' 'Bout Jesus" into "Talkin' 'Bout You." The marketers and journalists tasked with categorizing things called the music "soul," appropriately for its ability to stir the inner presence while setting the body into motion, and the jet-age buzz and snap of the electric organ were integral to its sound. The Hammond broke categories apart, with the keyboardist Jimmy Smith (see photo, page 144) recording soulful, danceable organ music for a jazz label, Blue Note, and making pop hits with albums like *The Sermon!*

By 1966, some fifty thousand churches in the United States had electric organs, and *Shindig!*, a prime-time pop music show on the ABC TV network, featured a soulful organist, Billy Preston, as a series regular showcased often in solo numbers, singing and playing a Hammond organ. Preston's most recent album, recorded shortly before his nineteenth birthday, was a collection of funky keyboard instrumentals titled *The Most Exciting Organ Ever*. He had been playing organ professionally since childhood in the early 1960s, when he toured with Little Richard, after Richard had abandoned rock 'n' roll to sing gospel music and then decided to return to rocking after all. Preston was performing with Richard in Hamburg in 1962 when he first met the Beatles. And that, in a circuitous way, led to his being invited eventually to play organ and electric piano with the Beatles themselves. On *Abbey Road*, the final album they recorded, Preston played electric organ on the John Lennon track "I Want You," though his playing is overwhelmed by the sound of the Moog IIIp synthesizer, which George Harrison had trucked from his house in Surrey to the EMI studios for the Beatles to experiment with.

All four of the Beatles noodled with the Moog in those sessions, and the synthesizer shows up conspicuously in tracks by each of them: Lennon's "I Want You," Harrison's "Here Comes the Sun," Paul McCartney's "Maxwell's Silver Hammer," and Ringo Starr's "Octopus's Garden." Lennon's idea was to use the machine to generate an insistent rush of white noise, which would grow stronger, ever louder in the mix, until it drowned out every other sound on the track. The mix the Beatles ended up using fell a bit short of that: some sounds other than the Moog noise are still discernible before the track ends up abruptly with a razor slash of the tape, but the effect of a mounting tsunami of extra-musical electronic sound came across, and there was nothing like it on any pop record of the time.

"We used the Moog synthesizer on the end [of 'I Want You']," Lennon said, as he was quoted in the Beatles' *Anthology* book. "The machine can do all sounds and all ranges of sounds—so if you're a dog, you could hear a lot more. It's like a robot. George can work it a bit, but it would take you all your life to learn the variations on it."

⌇

IN THE BATTLE OF WILLS BETWEEN HUMAN AND MACHINE, THE MOOG synthesizer won the early rounds. Then came Wendy Carlos.

To make the breakthrough Moog album, the first record of music produced entirely by synthesizer to hit the *Billboard* charts, Carlos had to customize components and parts she special-ordered from Robert Moog and devise her own arabesque system of techniques to accommodate the technology's abundant quirks. The project was a collection of pieces by Johann Sebastian Bach adapted to the Moog and performed by Carlos, released by Columbia Records late in 1968 under the title *Switched-On Bach*, a reference to the recording's electronic character with a wink in the direction of drug-savvy, turned-on young people who might ordinarily not buy albums of baroque music. Since the keyboard instruments of Bach's time could play more than one note at a time, Carlos had to work out a

way to record single tones then record over them multiple times in perfect synchronization, using an instrument that slipped out of tune easily and often. Working primarily in a studio she built in her one-room apartment in Manhattan, Carlos put more than a thousand hours into the project over five months' time.

Her objective was, essentially, to humanize the synthesizer—in her words, "to demonstrate to the world that electronic music did not equate with a stereotypically weird and unapproachable collection of disjunct bleeps and boops lacking anything one might call musical expression and performance values." A music scholar with a graduate degree from Columbia-Princeton Electronic Music Center, Carlos was a high-level thinker who gave human feeling primacy in music she made by means of electronic synthesis. She prized emotional truth in this and the larger body of synthesizer work she recorded after *Switched-On Bach*, a few titles of which were initially released under the name she was given at birth, Walter Carlos. (See photo, page 158.) During the period *Switched-On Bach* was being made, Carlos was undergoing hormone treatments for gender dysphoria.

The revered pianist Glenn Gould, a notorious miser with praise, admired the entirety of Carlos's achievement with *Switched-On Bach*. "The whole record, in fact, is one of the most startling achievements of the recording industry in this generation and certainly one of the great feats in the history of 'keyboard' performance," Gould said in an interview on Canadian radio. "Theoretically, the Moog can be encouraged to imitate virtually any instrumental sound known to man, and there are moments on this disc which sound very like an organ, a double bass or a clavichord, but its most conspicuous felicity is that, except when casting gentle aspersions on more familiar baroque instrumental archetypes, the performer shuns this kind of electronic exhibitionism."

Carlos, in an interview of her own with the same station, spoke humbly of her work and looked ahead to a time when, she predicted, synthesized music would become broadly accepted—normalized, no longer taken as a novelty. "We're just a baby," she said. "Although now we can see that the child is going to grow into a rather exciting adult, we've still got to take one step at a time. It will become assimilated. The gimmick value—thank God—is going to be lost, and true musical expression, and that alone, will result."

Switched-On Bach was an unexpected hit, a record people seemed to need, and not just want, to satisfy their curiosity and be attuned to the increasingly electrified world. Released late in 1968, it hit the top of the *Billboard* classical chart in January 1969, and stayed in the number-one position for the next three years. It simultaneously crossed over to the pop charts, where it made it to number 10, in between The Association's *Greatest Hits Vol. 1* and *Bayou Country* by Creedence Clearwater Revival. Before the end of summer in 1969, it was certified gold by the Recording Industry Association of America. Within five years, it would sell more than a million copies.

The critical response was tempered, as if critics needed time to come to terms with the synthesizer and its possibilities. The presence of Bach as the source of the music appeared to help some people accept the equipment Carlos used as part of a continuity with the past, with one writer explicitly linking the Moog to historical machines—that is, the machinery of musical instruments. "It would be easy to pass this recording off as just another gimmick of our ultra-gimmicky age," wrote the critic Thomas Willis in the *Chicago Tribune*. "Who cares if expressive and musically well-informed performances of Bach can be synthesized in the laboratory? There are plenty of other machines—violins, organs, flutes, and the like—which can produce the sounds more easily and economically. Operators for these, while not as plentiful as electronics engineers, are still available.

"Tomorrow, *Switched-On Bach* may be a collectors' item documenting today's first halting steps across another bridge between science and art."

Weird-looking and arcanely technical, an object of fascination to kit-building hobbyists and sci-fi nuts, the electronic synthesizer was generally conceived of as square, white, and male enough without being associated with music as square and white as Bach. Attitudes toward synthesizers began to shift, with Wendy Carlos helping to change the way perceptions of electronic music were gendered. Still, the synthesizer was far from being seen as the coolest, hippest thing around, until 1972.

As the preface to the story of this hard turn in attitudes toward the synthesizer, a twenty-eight-year-old audio engineer named Robert Margouleff heard *Switched-On Bach* when it was released, in 1968. Loosely associated with Andy Warhol's Factory, Margouleff had been making his living in advertising work while pitching in on Warhol projects and following his idiosyncratic curiosities. (Margouleff would later be co-producer of the Warhol film *Ciao! Manhattan*, a vérité semi-biography of Factory "superstar" Edie Sedgwick.) After buying a Series IIIc synthesizer from Robert Moog and befriending its inventor, Margouleff set out to build a mammoth super-synthesizer that could outperform the Moog. Working in collaboration with Malcolm Cecil, a British jazz bassist with a tech bent who had recently moved to New York, Margouleff cobbled together multiple Moog components with additional modules and several keyboards, concocting what turned out to be the largest, most advanced music synthesizer in the world, a machine capable of playing polyphonically with quasi-symphonic effects. Margouleff and Cecil called it The Original New Timbral Orchestra, or TONTO. "It's not one instrument," Margouleff would say. "It's all instruments at the same time."

Composing, experimenting, improvising, and having fun together, Margouleff and Cecil produced an album of original synthesizer music titled *Zero Time*, and it was released in 1971 with the artists credited as TONTO's Expanding Head Band, a name Margouleff thought of while tripping on peyote. The music captured that provenance: it was electronic head-trip music, spacey but melodic and oddly, unpredictably beautiful. As Mark Mothersbaugh of Devo would describe it in notes for a reissue of *Zero Time* on CD, "TONTO represented the cutting edge of artificial intelligence in the world of music. Robert and Malcolm [were] the mad chefs of aural cuisine with beefy tones and cheesy timbres, making brain chili for those brave enough and hungry enough."

An American bassist friend of Margouleff's, Ronny Blanco, was acquainted with Stevie Wonder and gave him a copy of *Zero Time*. On the afternoon of Saturday, May 29, 1971, early in Memorial Day weekend, Blanco showed up at the studio where Margouleff and Cecil had set up TONTO, accompanied by Wonder, who was carrying *Zero Time* under one arm. As Malcolm Cecil would recall, "He said, 'I don't believe all this was done on one instrument. Show me the instrument.'

"He was always talking about seeing," Cecil said. "So we dragged his hands all over the instrument, and he thought he'd never be able to play it." Within an hour, the three of them had started recording, with Wonder playing the multiple keyboards Margouleff and Cecil had set up. By Monday night, the end of the holiday weekend, Wonder, Margouleff, and Cecil had worked with the TONTO to record seventeen full tracks of new music, including all the selections that would soon appear on the album *Music of My Mind.*

The timing was not random. Just two weeks earlier, on May 13, 1971, Stevie Wonder had turned twenty-one, the date he had been awaiting for years to be released from his contract with Motown Records. Berry Gordy has signed the outrageously gifted Stevland Morris Judkins, in coordination with his parents, when he was just eleven years old, a minor who would remain a minor for ten more years. Gordy renamed him and was soon promoting him as Little Stevie Wonder, the 12 Year Old Genius. Working steadily at Motown, the hit factory Gordy ran on the model of the Ford assembly line, Wonder generated a string of chart-topping records made Berry's way: "Fingertips," "Uptight (Everything's Alright)," "I Was Made to Love Her," "Signed, Sealed, Delivered (I'm Yours)," and more.

"I was trapped for many years," Wonder would explain. "I've always felt I've been confined within a set style of work—that people expected a certain thing from me. Like, 'Stevie Wonder appeals to this—this is him.' I think a lot of artists are categorized or labeled in this way, and it's bad. People should let you be as free as possible and up until now I really haven't been."

Wonder saw liberation in the synthesizer's ability to do all the work Berry had had done by the Motown production staff. "Stevie was apparently quite taken by the idea that this was a keyboard instrument that he could possibly play that made all of these sounds," Cecil recalled. "He was tired of having to play his songs to an arranger who would then go away and write the arrangement, record the track with the band, call Stevie in after it was recorded, tell him where he had to sing, what he had to sing, and then send him away again while they did the mix. And Stevie said it sounded nothing like what the song sounded like in his head."

The expansive technology Margouleff and Cecil put in Wonder's hands stirred his creative imagination and seemed acutely suited to an artist who, having lost his sight soon after birth, had a highly developed sensitivity to nuances in sound. "I didn't know what he expected from it," Margouleff said. "I think Stevie acted on impulse."

Wonder was taken by the TONTO's ability to generate unorthodox and malleable sounds that felt more like abstractions, like thoughts in aural form, than the output of a musical instrument. He found using the synthesizer to be "a way to directly express what comes from your mind," as he told a reporter at the time. "It gives you so much of a sound in the broader sense. What you're actually doing with an oscillator is taking a sound and shaping it into whatever form you want. Maybe a year and a half ago I couldn't have done these kinds of tracks. I don't know. I think your surroundings and environment have a great deal to do with what comes out of you, how you write." Using a machine with no presets, Wonder was able to conceive and execute sounds never heard before. In fact, the early synth-tech limitations of TONTO required that he work that way.

Returning to Detroit, Wonder leveraged this new material of a whole new kind to negotiate a new contract with benefits unprecedented at Motown, including complete creative freedom to compose, perform, record, mix, master, and own all the music released under his name at a high royalty rate (reported to be between 14 and 20 percent of record income). Beginning with the appropriately titled *Music of My Mind*, released in 1972, Wonder would work with the TONTO to make four albums of technically radical, openly emotive, and wholly original music, repositioning its creator as a fully mature and singular creative force. The albums included *Talking Book*, *Innervisions*, and *Fulfillingness' First Finale*, with the singles "You Are the Sunshine of My Life" and "Superstition," his first number-one hits in a decade, "Higher Ground," "Living for the City," "You Haven't Done Nothin'," and "Boogie On Reggae Woman," among others.

On Wonder's biting new songs of social critique, especially the grim operetta "Living for the City," the eerie programming of the TONTO darkened the atmosphere and heightened the tension. On rhythmically propulsively songs like "Superstition" and "Higher Ground," the piercing synth tonalities brought a dimension of unsettling mystery to Wonder's virtuosic, churning keyboard work. The weirdness of synthesized sound found a way out of the tech lab and onto the dance floor. Your body's grooving, and your mind is wondering, *What the fuck is that sound?* "I love getting into just as much weird shit as possible," Wonder told the writer Ben Fong-Torres in an interview about his first synthesizer records for *Rolling Stone.* The TONTO provided him with sonorities, textures, layering options, and rhythmic possibilities as weird as musical shit could get in the 1970s. The music would stand as the high-water mark of Wonder's career and establish him as prince of a new domain of synth-laden, Afro-futurist pop. "This collaboration [between Wonder and the creators of TONTO] changed the perspectives of Black pop music as much as the Beatles' *Sgt. Pepper's Lonely Hearts Club Band* altered the concept of white rock," wrote the critic John Diliberto.

As synthesizer technology progressed, Stevie Wonder would go on to compose and record fruitfully, with unabated imagination and no further need for the TONTO, although tracks he had made with it appeared on several of his later releases. At the same time, Margouleff and Cecil worked with a range of other artists of color impressed by Stevie Wonder's work with the TONTO—Gil Scott-Heron, the Isley Brothers, Quincy Jones, Wilson Pickett, Billy Preston, Weather Report, Bobby Womack, and others—to embed the sound of the synthesizer into Black pop of the late twentieth century.

"Just look at what Stevie did with the synthesizer, and then what I did when I got my hands on it," Quincy Jones told me. "It's a machine for making sculpture with sound—for sculpting sound. It takes a pure electronic signal and sculpts it into something beautiful. Or it can, if you know how to work with it, like I do. Don't you dare try to denigrate the value of technology for making music—not to me. I know better.

"The thing is, the thing is this: People didn't expect all that innovation to come from Black musicians in the pop arena. They forgot, nobody knows how to take something that looks like it's nothing and then *prove to you* that it's really something the way a Black artist can. One thing a Black man understands is that something may look like it's nothing, but that doesn't mean it's nothing. I'm talking about a man, but that's not all I'm talking about."

12

PARADISE

B y the last quarter of the twentieth century, the "man–machine" concept was so timeworn that it could be applied only in historical reference or in irony. Rhetorically, the phrase was doubly outdated for both its male orientation and its suggestion of the physically mechanical in the era of electronics. The word "machine" would survive well into the twenty-first century, when machine learning would be established as the methodology of artificial intelligence. Still, in 1977, when the German art-rock band Kraftwerk recorded an album titled *The Man-Machine*, the phrase was employed as tactically retro modernism, conjuring images of the fantastical Industrial Age utopianism of *Metropolis* and the Italian Futurists—the ethos Jane Heap had sought to institutionalize with the Machine Age Exposition of 1927. Kraftwerk exploited the term for its evocation of prewar history, while repurposing it with irony as a catchphrase for electronic-era fetishism.

Technology had always been integral to Kraftwerk's art. Founding members Ralf Hütter and Florian Schneider, a keyboard player and a multi-instrumentalist respectively, met in music school and played together originally in an experimental rock group called the Organisation, signaling their early interest in order and discipline. Recasting the group with a few different musicians in 1970, Hütter and Schneider renamed it with the German word for power plant, a term that in their time and place would trigger images of nuclear energy facilities like the Rheinsberg nuclear power plant and the Greifswald nuclear power plant in East Germany. If Americans listening to the Eagles heard about this group from Europe and wondered what it had to do with "craftwork" like crocheting and woodcarving, Germans were primed to relate Kraftwerk to the wonders of Atomic Age science and the dark possibilities beneath its shimmery surface.

The cover of the *Man-Machine* LP showed the four members of Kraftwerk dressed in primary-red dress shirts and thin black ties, standing in rigid formation at a three-quarter angle. It was designed in homage to the Russian constructivist artist El Lissitzky, whose use of mechanical angularity and a stark red and black color scheme had informed the aesthetics of the Bauhaus and De Stijl art movements in prewar Europe. The Kraftwerk musicians wore the short, brushed-back haircuts of military cadets—a subversive look in the day of mullets, shags, and big-hair guitar bands—and their lips were painted in bright red to match their shirts. Packaged in visual codes of proto-retro-futurism, the *Man-Machine* album contained six selections of tightly wrought music performed mainly on pre-digital electronic and electrified instruments, including a portable synthesizer introduced in the early Seventies, the Mini-Moog, as well as its knockoff, the ARP Odyssey desktop synth. The rhythms were electronically generated and sounded proudly, mechanistically so; and the singers' voices were heavily processed to resemble those of drive-in movie robots. The songs—"Metropolis," "Skylab," "The Man-Machine," and three more—intermixed lyrical and musical suggestions of old Berlin music halls and life in space: music of the Weimar Jetsons.

As Hütter described it, the making of the album was a collaboration between the men and their machines. "We are playing the machines, the machines play us," he said. "It is really the exchange and the friendship we have with the musical machines which make us build a new music."

The late 1970s was a time of vital musical ferment in what was then West Germany, with emerging technologies helping to redefine and reinvigorate a landscape still divided into disparate spheres under cultural occupation by the competing East and West forces of Cold War geopolitics. In the same year Kraftwerk recorded *The Man-Machine*, David Bowie released a pair of synthesizer-laden albums, *Low* and *Heroes*, both made with the futurist Brian Eno and recorded primarily in Berlin. (Along with the more eclectic *Lodger*, released two years later, the projects would become known as the Berlin Trilogy and be widely regarded as among Bowie's best and perhaps his all-time best three albums.) While he was working in Berlin, Bowie also arranged for his American friend Iggy Pop to join him in Europe, so Bowie could help update Iggy's sound with injections of electronics and mechanical rhythms. Using many of the same musicians who were playing on his German projects, Bowie produced what would be Iggy Pop's first solo album, *The Idiot*. (All this is indirectly related, but nonetheless related to the conception of mechanical noise as art music that led Lou Reed, a friend of both Bowie and Iggy, to make *Metal Machine Music*, a notoriously listen-proof double album of severe electric guitar drone sounds released in 1975 and pulled off the market three weeks later.)

The Idiot concluded much the same way Iggy's life in music had begun. On the closing track, "Mass Production," a frightening/exciting synthesizer drone overtakes the sound of the musicians playing detuned electric guitars. Bowie and Iggy had synthesized an atmosphere of alluring mechanical havoc to match the industrial menace that the young Jim Osterberg had found exhilarating at the River Rouge industrial complex. Iggy would describe *The Idiot* as "a cross between James Brown and Kraftwerk."

Five hundred kilometers to the south, in Munich, Giorgio Moroder, a thirty-four-year-old composer and guitarist, was experimenting with ways to use the Moog synthesizer to make pop records. Born in northern Italy, Moroder was now living in West Germany and making his living as a recording studio engineer. "That was a very interesting moment for me," Moroder told me. "I heard the recording of Walter Carlos [sic] playing the Moog synthesizer—*Switched-On Bach*—and I really liked it. I *loved* it! I felt like Carlos got inside me and found the real me. I told myself, 'This is what I want to do! This is *me*!'"

Through inquiries to contacts he had made in the German music scene, Moroder found a composer in Munich, Eberhard Schoener, who was using a Moog to make avant-garde *musique concrète*. Schoener let Moroder experiment with the synthesizer, and "that was it," Moroder said. "That was the beginning, and it would never end for me. I sat down and studied it and learned everything—all the sequencers, everything, everything every control could do. I thought, 'This is my world. I gave myself up to the synthesizer. I said, 'Take me now. Show me what you can do.'

"I didn't want to play Bach or make *musique concrète* like Schoener," Moroder explained. "I wanted to make commercial music—popular records for everybody, but sexy records, popular and sexy. And that's what I would do, using just the synthesizer. I didn't need other people. All I needed was the synthesizer and a singer, a sexy woman singer."

Moroder first met the sort of singer he had in mind about two years earlier, when Peter Bellotte, an English producer, introduced him to an American vocalist Bellotte had worked with on a couple of singles. She had come to West Germany as Donna Gaines in a touring production of *Hair*, liked the country, learned the language, married an Austrian actor named Helmuth Sommer, had a baby, and settled in Munich. Building upon a song idea of hers, Moroder and Bellotte devised a sultry, throbbing, partly electronic dance-music track. With the addition of vocals—including an audacious section of moans working up to a simulated orgasm that proved the singer's exceptional ability to embody a song—the track was released as "Love to Love You" (with a final word, "Baby," added to the title on later pressings) under the name Donna Summer. An international hit, the song became an anthem of the disco movement as it migrated from the underground gay clubs of Greenwich Village to the pop charts and the gym floors of high school dances everywhere.

"We had a very nice hit with 'Love to Love You,' and we wanted to follow it with something bigger and better, more fabulous," Moroder would recall. "Between the records, I started to understand really what synthesizers could do and how they could move things ahead."

Moroder and Bellotte were collaborating again, with a conceptual idea for what would be Donna Summer's fifth album, to be called *I Remember Yesterday*. (At first, the title was to be *A Dance to the Music of*

Time, after Anthony Powell's twelve-book series of ruminations on cultural history, which Bellotte said he had recently finished reading.) Each song would mirror the musical sensibility of a stage in pop history, from the swing era of the midcentury through the Brill Building girl groups of the early Sixties to contemporary disco. But all the tracks of historical pastiche would also have a Seventies dance-music feeling. The final selection would look beyond yesterday to tomorrow, conjuring the musical future—a gaze into the disco crystal ball. For that last piece, Moroder arranged to borrow the composer Eberhard Schoener's Moog IIIp modular synthesizer, and he hired Schoener's studio assistant, Robert Wedel, to help with the equipment.

"We wrote it backwards," Moroder said, recalling the gestation of the album's final selection, which ended up with the title "I Feel Love." He and Bellotte constructed the instrumental track with no ideas for the melody or the lyrics, which Donna Summer later created and overdubbed. Musicians of the twenty-first century would recognize this process as beat-making and top-lining, the practice by which virtually every pop recording is made today. "We followed the synthesizers," Moroder explained. "What can they do? Where can they take us? We followed the technology."

The instrumental track (beat) was unusually spare and made almost entirely electronically. The synthesizer provided an infectious blip-blip-blippy groove with unchanging mechanical precision, as well as a synthetic bass line in a different tonality and a simulation of the ching of a hi-hat cymbal. From time to time, cool swooshes of electronic tones washed in and out. The only sound produced by an acoustic instrument was a simple kick-drum pattern. (Ever since its release, many writers would mistakenly describe "I Feel Love" as being created entirely electronically, and I made that error in print myself once.) This was music pared down to a handful of components, nearly all of them boldly electronic in character, with everything synchronized in lockstep. To achieve the precision linking of synthesizer parts during the making of the track, Wedel proposed synching the elements with a common signal, a procedure Moroder and Bellotte had never heard of before. Robert Moog knew nothing of it, either. The technique was Wedel's invention and another one people recording in the twenty-first century would recognize. Virtually all music produced digitally today is synced to "the click" during performance in the studio or synced to "the grid" on editing software during the mixing process.

As the writer Richard Vine would later point out in the *Guardian*, "I Feel Love" was "one of the first [recordings] to fully utilize the potential of electronics, replacing lush disco orchestration with the hypnotic precision of machines." The record used electronics, an internal kind of machinery—the machines that mirror the human mind—to prod the body into movement.

When the record was released in July 1977, David Bowie was still working on *Heroes* at the Hansa recording facility in Berlin. Brian Eno burst into the studio with a copy of the single in his hands, and waved it in Bowie's direction. "This is it—look no further!" Eno announced. He had discovered "the sound of the future."

Giorgio Moroder had done something more than imagine the future for a concept album of pop-timeline pastiche. Responding to the sounds the synthesizers made, he carried pop and dance music into a future that he and the technology were inventing. "If any one song can be pinpointed as where the 1980s began, it's 'I Feel Love,'" wrote the critic and historian Simon Reynolds, author of the definitive critical study of Eighties dance music and rave culture, *Generation Ecstasy*.

Robert Moog, upon hearing "I Feel Love," recognized the uniqueness of the recording, but thought of what made it unique as a failing. "The sequencer bass that's chugging along through the whole thing has a certain energy to it but also a certain sterility, because it's always the same," Moog said, complaining of this effect on both "I Feel Love" and a song the jazz-rock guitarist Jeff Beck made with a Moog.

Sterility and sameness. Coldness and uniformity, detachment and regimentation, these were long-familiar criticisms of art or music made with or by machines, be they mechanical or electronic, analog or digital. There is no question that the rigidly exact, unvarying rhythms produced by synthesizer programming on "I Feel Love" gave the music a quality of mechanized sameness that feels nonhuman. One has to wonder, then: Why did so many people love this music and feel compelled to dance to it? How could a mechanical quality and sameness be assets, not liabilities?

THERE ARE GOOD REASONS FOR HUMAN BEINGS TO RESIST IDENTIFICATION
with machines of any kind: machinery is incapable of feeling, and it has
no spirit—no essence that transcends the physical. It lacks the qualities we
tend to prize most about our identity as living beings and our place in the
grand scheme, as well as we can understand such things. By extension,
we're disinclined to look favorably on ways of living and modes of acting
connected to machinery or industry. We reject regimentation, and, if we
can, avoid dehumanizing labor that reduces us to serving as anonymous
elements in an industrial scheme—like Chaplin's factory worker in *Modern Times*, as I noted earlier. (Parallel jobs are still plentiful, with brigades
of workers shuffling boxes in Amazon distribution centers.) We don't live
in warehouses, factories, or garages; we don't want to be machines.

And yet, there was a period in the second half of the twentieth century
when a significant number of people united by common bonds of identity
took action together in ways long associated with machinery, subverting
the tropes of dehumanizing industry to assert their humanity. Gay men
of color, in particular—along with other people who did not conform to

traditional sex and gender categories—gathered late at night in underground locations such as an out-of-business truck garage, an abandoned warehouse, and other disused industrial sites whose provenance in industry was left undisguised. Amassing by the hundreds and the thousands, they danced with their bodies in synchronized motion, jacking to a new kind of music stripped of all niceties to the raw, bare sound of hard-pounding rhythms and electronic noise. They moved in unity with the precision of machine parts and kept moving, nonstop, for hours, and all for a purpose that transcended industry and commerce.

The music that drove them grew directly from Kraftwerk and Giorgio Moroder's work on "I Feel Love"—man–machine dance music, constructed with technology in styles that would become internationally known as house music and techno. It was created in nontraditional ways by an unprecedented class of music makers who were not playing musical instruments. They were using machines to produce sounds to stir people to move like components in a machine—a machine of social transformation.

One of the most notable early innovators in this phenomenon was the DJ Larry Levan, a wiry, high-spirited, young, occasionally orange-haired Black man from Brooklyn. A former acolyte at an Episcopal church in Brooklyn, Levan picked up the rudiments of mixing live audio on the church sound system as a boy. Though he dropped out of high school, he took courses in textile design at the Fashion Institute of Technology, where he met another gay Black music lover, Frank Nicholls. A former altar boy at a Catholic church in his native Bronx, Nicholls was a striking visual complement to Levan: stocky and bearded, an imposing figure who took up a street name more befitting a mob *caporegime*, Frankie Knuckles. Together, they found their way into New York's ball culture as dressmakers for drag acts, and migrated from there to the Gallery, a gay disco in Chelsea, doing interior decoration. At the Gallery in the early 1970s, house DJ Nicky Siano showed Levan and Knuckles how he used three turntables to keep music spinning in a constant flow, demonstrating the craft with records of R & B, smooth-groove jazz, and the occasional danceable rock track.

The skill Nicky Siano tutored Levan and Knuckles in, DJing, was most overtly one of connoisseurship and curation: applying a nurtured command over a vast body of materials—records made in multiple genres and styles, including obscurities and the more obscure the better—to present in sequences both responsive to and stimulating to the dancers at a given time and place. Great DJing called for the having of great taste for the purpose of taste-*making*. In this, the DJ upended the assumptions in an old joke musicians told about would-be musicians they considered non-musicians: "What does he play?" "The record player."

Beyond the estimable task of curating music for a body of people with refined tastes of their own and little tolerance for the passé or the ill-fitting, DJing demanded proficiency at turntable technique. Indeed, the complex set of ways DJs manipulated turntables not merely to play music but to remake the music turned the DJs into musical creators. *What does he play?* For DJs in both club music and hip-hop, to say "the turntable" would be to signal mastery in an emerging artform. A machine for playback had become reconceived as an instrument for reinvention.

DJs in hip-hop, in particular, were in the process of innovating an original sonic vocabulary constructed through turntable manipulation. DJs learned to isolate the break—the brief part of a dance track where most of the musicians take a break and leave the drums (or drums and bass) to carry the rhythm, for dramatic effect—and extend it by mixing breaks with matching tempos from different records on multiple turntables. They worked out how to punch excerpts of music from one turntable onto the break played on another. Early in the development of turntabling as a creative art, DJs such as the hip-hop innovator Grand Wizzard Theodore devised techniques to exploit the physicality of vinyl records to make new sounds, or to build upon and reapply old sounds like the scratching noise the needle produces when the record spins the "wrong" way. If musical instruments are machines, and nearly all are, DJs proved resourcefully inventive at turning machines into musical instruments.

With the skills in club DJing he picked up from Nicky Siano, Larry Levan talked himself into a job in the tech booth at the Continental Baths—at first, working the lights, then doing fill-in DJ work—and Knuckles followed him there. A luxuriously appointed complex in the basement level of the Ansonia Hotel on the Upper West Side of Manhattan, the Continental Baths had a long pool for wading, swimming, and having sex; a lounge for unwinding over cocktails and having sex; a ballroom area for dancing and having sex; and private rooms for napping between all the sex. There were some live performances—typically, by young performers between above-ground jobs, like the Broadway understudy Bette Midler and her pianist, Barry Manilow. The primary entertainment, though, was the nonstop music provided by the DJs: Knuckles doing the warm-up, starting around midnight, then Levan though the night to the morning hours. Working the Baths, Knuckles began to develop a signature approach to curation, building mixes of richly textured Philly soul tracks and disco, while Levan worked up a diva act behind the board, miming and swooning to the music.

By 1977 (that year again), Larry Levan was a rising DJ star invited to be the main act at a new club, the Paradise Garage, opening in an abandoned parking garage on the industrial fringe of the West Village—a rusting area with little residential housing, taken up by gay men and others with fluid, open, or socially transgressive approaches to identity. In the years between the Stonewall uprising in June 1969 and the first reports of a mysterious contagion not yet identified as HIV/AIDS in 1981, the community that would one day be known as LGBTQ was actively engaged in becoming a community, and much of that activity was taking place on the dance floors of clubs like the Paradise Garage—spaces where people who had long been forced to hide could come together as a body and take pleasure in one another's company. (See photo, page 173.) From that pleasure grew pride. At the Garage, some 1,400 people per night could dance together, expressing in unison movements to persistent beats that they were united and had the strength to persevere. (And 1,400 was merely the legal capacity of the space. On busy nights, which by the end of the Seventies meant most nights of the week, the club was stuffed beyond capacity, with shirtless, sweating bodies of gay men, the majority of them men of color, pressed against one another.)

The music propelling them boomed at rocket-blast force from a custom-built sound system, with Larry Levan spinning the tracks, seamlessly fading one into the next: James Brown's "Give It Up or Turnit a Loose" into Kraftwerk's "The Robots" into Loleatta Holloway's "Hit and Run" into Donna Summer's "I Feel Love." Levan soon learned how to edit tracks onto open-reel tape and extend the breaks, so what were once short transitional passages of drums and bass became songs of their own—a new kind of song without song structure, the music reduced to its rhythmic core. He was doing the same kind of thing that pioneering hip-hop DJs were doing in the Bronx, but with no one rapping and adding words to complement and complicate the music.

"At the Garage, I felt like I was a part of something more powerful than myself," wrote Christopher Vaughn, who worked in marketing for General Electric. "At the office, I was the only Black man and one of a tiny handful of gay people. I felt powerless. At night, with the music blasting through us, every one of us there felt empowered. In my job, I would sit through meetings with engineers explaining how electronic technology worked. I wanted to say, 'Honey, *you don't know*! Come with me, and I'll show you what it's like to be *part of a machine*.'" As a social mechanism for gay Black men like Vaughn, house music was a force of strength and power, in part because the dance floor felt like a literal mechanism.

By the end of 1977, a new dance venue was about to open in Chicago, and the owner offered Larry Levan a position as the main DJ. He declined, reveling in his budding club-superstar status in New York. Levan recommended Frankie Knuckles, who took the job and moved out of New York.

The venue was a vacated storage warehouse in a bleak area of post-industrial Chicago. Though the official name of the club was US Studio, everyone called it the Warehouse and that got shortened to the House. The dance floor was in the basement level of a four-story, yellow-brick building on South Jefferson Street, an area spotted with a few blue-collar operations still functioning by day. "People would say it was like climbing down into the pit of hell," Knuckles recalled. "People would be afraid when they heard the sound thumping through and saw the number of bodies in there, just completely locked into the music."

Living in a city where he was unknown, working in a club with no history, Knuckles felt free to experiment. He intermingled styles of music, shifting from soul to Euro synth-pop to disco, with a special affection for the seemingly incompatible sounds of Philadelphia soul and Kraftwerk. As he explained, "Kraftwerk were main components in the creation of house music in Chicago." He combined "Philly sound with the electro beats of Kraftwerk and the electronic body music bands of Europe."

Manipulating the materials, often working at home to prepare tapes for playing at the club, Knuckles smoothed out the varied source sounds to a danceable flow by altering the tempos of some tracks and, significantly, layering in mechanical beats from a drum machine. He would often boost the bass, sometimes editing in a basic, thumping sequence of low-frequency tones with a mini-synthesizer. By the early Eighties, Knuckles was essentially producing new musical works from found materials, and they had a unified character: with fancy complexities of instrumentation, melody, and harmony subordinated in the mix or removed in the edit, the music had a mechanized purity. It had an abstracted, inhuman perfection.

"The kids that were hanging out at the Warehouse . . . I didn't know who they were," Knuckles remembered. "And they started having different parties on their own in these different taverns and bars in Chicago. And when they'd do this, they had a lot of success with it. And one day, I was going out south . . . and on the corner there was a tavern, and in the window it had a sign that said 'We play house music.'

"I asked this friend of mine, 'Now what is that all about?' And she says, 'It means music like you play at the Warehouse.'"

The authors Bill Brewster and Frank Broughton, in their book *Last Night a DJ Saved My Life: The History of the Disc Jockey*, offer an etymology of the phrase "house music" independent of the Warehouse:

> If a song was "house," it was music from a cool club, it was underground, it was something you'd never hear on the radio. In Chicago, the right club would be "house," and if you went there, you'd be "house," and so would your friends.

In that sense, the term is an echo of the social system under which ancestors of today's LGBTQ community lived in the early twentieth century, when homosexuality was a criminal offense in most of the United States. People leading socially transgressive lives mainly congregated in secret parties in private homes, the safest place for them to meet. When they played naughty songs like "He's So Unusual," "BD Woman's Blues," and "The King's a Queen at Heart" on their Victrolas to dance together, it was music made to share only where they lived: house music.

An unorthodox art constructed by electronic means, house was a defiant outlier in musical terms. Like ragtime on piano rolls during the first craze for Black music in America, house was a radical art produced by mechanical means. By traditional standards, it was "wrong" for its relentless repetition, its severe synthetic tones, and the way it broke most of the rules of harmony and song structure. It was an outcast music for a people coming together to defy a history of being cast out. It was a harbinger of changes to come—musical as well as social and political changes—with house music laying the groundwork for rave culture and the cross-cultural electronic dance music of the late twentieth and early twenty-first centuries.

Unlike hip-hop—whose music was also constructed electronically through turntabling, mixing, and sampling—house was a right-brain art: emotional and extra-literal, with gestural lyrics that often amounted to no more than a catchy phrase repeated and repeated. In house music, no premium was put on linguistic dexterity, and no attention paid to the depth of the lyrical content. House was an artform intended not to stimulate the brain but to turn it off, so the body could take over. That house's core fans were gay or otherwise nontraditional in gender or sexual orientation largely isolated the audiences for house and hip-hop, the two new forms of Black music most indebted to electronics. "Rap and house were polar opposites, because rap people often found club music to be too effeminate, too sissy for them," said Brian Chin, a music industry executive quoted in the *Los Angeles Times*.

"My white coworkers at GE liked Bruce Springsteen. My Black friends from high school liked hip-hop," wrote Christopher Vaughn. "I liked house music—and Stephen Sondheim."

The critic John Leland described house music pithily in an early essay of appreciation. "House records are like 12-inch remixes with the original songs removed: All you hear are the inflated beats, the souped-up bass, and the frilly sound effects. Cheaply made, usually on local independent labels, they can be boring at home. But in the loud, crowded spaces of a club, they become intoxicating. With their huge beats and minimal arrangements, they provide the sense of space that the clubs don't; they are as much architecture as music.

"House is disco for a time when pleasure has become complicated: It is darker, more sinister, more fascinated by the cold dispassion of electronics."

By 1983, the success of the Warehouse was beginning to undermine its character. Growing numbers of straight and white men and women were going to the club as affection for house music became known as a mark of elevated taste, and Frankie Knuckles moved on. "The parties were very intense—they were always intense—but the feeling that was going on, I think, was very pure," Knuckles explained. "And a lot of that changed between '82 and '83, which is why I left there. There was a lot more hard-edged straight kids that were trying to infiltrate what was going on there, and for the most part they didn't have any respect for what was going on." Knuckles opened a new club in Chicago to carry on house music for gay, Black dancers. He named it the Power Plant, the English translation of Kraftwerk.

In need of some equipment, Knuckles arranged to buy a Roland model TR-909 drum machine from a fellow DJ from the Detroit area, Derrick May, who came to Chicago on occasion and made himself known at the Warehouse. "So he came down that weekend, and he brought it," Knuckles recalled. "I would use it live in the club. I would program different patterns into it throughout the week, and then use it throughout the course of a night, running it live, depending on the song and playing it underneath, or using it to segue between some things." The sound of the drum machine—unmistakably artificial and inhumanly exact—was more machine than drum, and it grew in importance to dance music as the music moved further into electronic mechanization. Easy to operate, economical, and undisguisably artificial, the drum machine became a staple of the techy pop that came to supplant baby boomer rock on Top 40 radio. A wave of young drummers like Chris Frantz of Talking Heads would be playing traditional drum kits but emulating the unchanging mechanical beats of drum machines.

Derrick May was playing an important role in taking club music further into purer mechanization, working in coordination with two other DJs, his friends Juan Atkins and Kevin Saunderson. They had all met a few years earlier, when they were students at Belleville High School, Black teenagers in a predominantly white suburban environment. Belleville was only seven miles from Ypsilanti, where Jim Osterberg had grown up. In fact, Juan Atkins attended the community college in Ypsilanti after Osterberg moved away from the town. Like the future Iggy Pop, Derrick May was influenced by early exposure to the noisy, grimy atmosphere of the factory works surrounding Detroit. Unlike Iggy, however, May found no beauty in the industrial environment. He found beauty in a different kind of machinery that stood in contrast to Detroit industry: the crystalline mechanical electronica of Kraftwerk and European dance music.

"It was just classy and clean, and to us it was beautiful, like outer space," May said. "Living around Detroit, there was so little beauty. . . . Everything is an ugly mess in Detroit, and so we were attracted to this music. It ignited our imagination."

May, Atkins, and Saunderson—the Belleville Three, as they would come to be known in dance-music circles—worked with drum machines, sequencers, and portable synthesizers to push the innovations of house, stripping tracks to the barest of essentials and processing them to make night-long mixes that did not quite mix the tonalities but slowly stirred them. Very, very slowly. Created to play continuously for hours at a club, the music throbbed with an unchanging mechanical pulse and minutely subtle shifts in sonic timbre rather than harmonic movement. It was an art of seductive constancy, designed to lead dancers into a state of unconscious movement and trance-like euphoria, and it worked as both a complement and stimulus to Ecstasy, the pleasure drug most favored in the clubs of the Eighties. (Cocaine probably ran a close second, I think, though there's a lack of reliable data on this.)

"Within the last five years or so," Juan Atkins said in 1988, "the Detroit underground has been experimenting with technology, stretching it rather than simply using it. As the price of sequencers and synthesizers has dropped, so the experimentation has become more intense. Basically, we're tired of hearing about being in love or falling out, tired of the R & B system, so a new progressive sound has emerged. We call it techno."

Atkins called it "post-soul" music, distancing techno from Motown and the historic modes of Black music associated with the Motor City. Techno, he said, was "music that sounds like technology, and not technology that sounds like music, meaning that most of the music you listen to is made with technology, whether you know it or not. But with techno music, you know it."

~~~

BACK IN GERMANY, KRAFTWERK WAS USING A CONSPICUOUSLY RETRO FORM of technology to confuse the matter of who or what was making its music. For part of its concerts, the band would be replaced by four life-size mechanized mannequins with faces molded to resemble those of the musicians. They were attired in the same stage costumes the band wore, but without sleeves, revealing robotic arms with the gears and wiring exposed. (See photo, page 166.)

At a press conference in 1983, Ralf Hütter defended the group's mechanical stand-ins with futurist aplomb. "One day the robots will be the ones who will answer your questions," he predicted. "They will have an electric brain and memories with all the possible questions. To get the answers, you will only have to press a button."

13

A VERY CURIOUS
RELATIONSHIP

The Swiss watchmaker Pierre Jaquet-Droz taught his son Henri-Louis the art of fine machine craft. Under the tutelage of the master, the apprentice learned to make precision timepieces and organ-playing dolls through observation and practice. As the skills of the younger Jaquet-Droz improved, he advanced from serving as an assistant to acting as a full collaborator, developing techniques of his own along the way. Eventually, they both knew, the apprentice would replace the master. That was how the long-established (and often romanticized) tradition of apprenticeship worked in the eighteenth century, and it was not far removed from the way computer artists and their programs would work together two hundred years later in the gestating days of a new tradition for making digital art and music.

In the early 1970s, the British painter Harold Cohen wrote one of the most important early computer programs for making visual art, and he gave it a human name, Aaron. He said he thought of it as his apprentice—as an assistant he compared explicitly to the studio aides whom fine artists like

Leonardo and Michelangelo employed during the Renaissance. In time, Cohen would refer to Aaron as his collaborator, an electronic peer in the art-making process. "Aaron is autonomous to the degree that it is not going to ask me to hold its hand or make a decision for it," Cohen explained to a journalist in 1995. He would make this point often, stressing the creative autonomy he had built into Aaron's makeup—a limited autonomy and one its programmer created, though a kind of freedom, nonetheless. "It does things that I would never have done as human artist," Cohen told another writer in 1997. When Cohen died, in 2016, Aaron was still functional. Over the years after that, Aaron, though dormant, would remain capable of producing new art that Cohen would never have done and now could never see.

Harold Cohen came to computers relatively late for an innovator in a field then populated largely by college students and recovered former students with highly charged frontal cortexes, a generation of venturesome brainiacs whose optimism and ambition were unaffected by much adult-world experience. Cohen was forty years old and a successful painter, a respected contemporary of British artists such as David Hockney and Bridget Riley, when, in 1968, he was first exposed to computer art at an exhibition called "Cybernetic Serendipity" at the Institute of Contemporary Arts in London. (The location was just a couple of blocks from the site of Egyptian Hall, where Maskelyne and Cooke once presented their nightly shows of magic and automata.) By the late Sixties, Cohen had had multiple exhibitions of his densely structured, semi-representational/mostly abstract canvases, and, along with his younger brother, Bernard, was one of the "Five Young British Artists" selected to represent Great Britain at the Venice Biennale in 1966. A forward-looking artist working in a medium centuries old, Cohen made paintings with imagery that sometimes puzzled the London art establishment. As the critic Nigel Gosling described one canvas of Cohen's—or attempted to and gave up trying—it "defies description, and that's its strength. It is working completely [on] its terms. You can use words like lyrical or logical; but to justify them, only the sight of the pictures will serve."

"Cybernetic Serendipity," a show of more than three hundred artworks and objects, was one of three major museum exhibitions dedicated to the convergence of art and technology in the same season. It ran from

August 2 to October 20, 1968, closing just one month before the opening of the Museum of Modern Art show, "The Machine as Seen at the End of the Mechanical Age," and the Brooklyn Museum's simultaneous exhibition, "Some More Beginnings." (It was in October of that year, too, that Sol Lewitt first introduced what would become a long series of conceptual artworks he called *Wall Drawings*, for which he provided detailed mathematical instructions that others executed on walls, a three-dimensional precursor to algorithmic art.) The idea for the London show came from Max Bense, the intellectual guru of Frieder Nake and other young European artists and scholars drawn to technology in this period of explosive action and interaction involving tech. Bense gave it a name meant to focus attention on cross-disciplinary work, drawing from the definition of cybernetics as the science of circularity: how varying forces from multiple spheres of existence—ocean currents, wind, and sailing ships, for instance, or commerce, aesthetics, and science—can interact as a system. Curated by the British critic and scholar Jasia Reichardt, the exhibition was an elegantly presented assemblage of experiments in computer use and electromechanical hybridization. "Cybernetic Serendipity deals with possibilities rather than achievements, and in this sense it is prematurely optimistic," Reichardt announced in her comments to open the show. "There are no heroic claims to be made, because computers have so far neither revolutionized music, nor art, nor poetry, the same way that they have revolutionized science."

As MoMA had, "Cybernetic Serendipity" displayed one of Jean Tinguely's Meta-Matic art-producing machines and an assortment of other eye-catching, buzzing, and flashing thingies, along with a collection of two-dimensional works on paper with images generated by computers and printed with graphic plotters. There was work by a British philosophy professor, Desmond Paul Henry, who had adapted the analog computer systems used in the bombsights of Royal Air Force planes to make a geometric drawing machine. There was art by Nake, graphical work by a Boeing engineer, some by a team of artists and scientists from Japan, and pieces in a similar mode by others in the UK and Western Europe. Critics, once again, had some difficulty with art that was not made solely by artists, but created in collaboration with electronic machines. In an

Associated Press report on the show published nationally in the U.S., the critic Barbara Gold made a point of the limitations of the computer in art-making, while allowing that it could make contributions its human collaborators had not thought of. "It cannot think," Gold wrote, laying out a premise some computer scientists of 1968 would dispute. "It could and did go from there, however, to make a seemingly infinite series of 'creations'—combinations of symbols, signals, images, numbers, or letters—from the data it contained. Man was the actual creator. The computer simply let him escape the labor involved in making the art, and it revealed possibilities he may not have extrapolated from his initial ideas."

Harold Cohen was not impressed. He left the show thinking that "either computers were awfully stupid, or everybody was doing awfully stupid things with them," as the author Pamela McCorduck recounted the moment in her book *Aaron's Code: Meta-Art, Artificial Intelligence, and the Work of Harold Cohen*. "Cybernetic Serendipity" got him thinking about computers, though, implanting an inchoate sense of the possibilities that the exhibition's curator had spoken of. The show was serendipitous. Cohen left England that autumn to teach painting in the department of visual arts at the University of California, San Diego, where he met a master's student in music, Jef Raskin. Having earned a previous graduate degree in computer science from Penn State, Raskin had arranged to teach in visual arts while studying in the music department. Raskin, who would go on to be one of the earliest employees of Apple Computer, contributing to the development of the Macintosh, was a hard-driven iconoclast, the kind of scholar who could take master's degrees in multiple subjects and teach undergraduates while tutoring private students, with no time to waste on non-necessities like an additional "f" in his first name.

Cohen, who wore oval wire-rim glasses and had a beard specked with gray, came across as another of the aging-hippie professors rampant on California campuses, until he spoke. He retained the crisp, clipped English speech of his upbringing, and chose his words with precision. When he joked, as he liked to do, the humor could be so subtle listeners might never pick up on it.

Cohen asked Raskin to show him how to program a computer, and Raskin briefed him on the principles and handed him an instruction manual to the Fortran programming language. Working mainly on his own for a period somewhere between six months and a year (his accounts vary), Cohen learned well enough to start writing his own program for making art. To make further progress on it, he applied for and was granted a visiting scholar residency at the Stanford Artificial Intelligence Laboratory (SAIL) that John McCarthy established after moving from Dartmouth. Cohen finished the program there and named it Aaron, using the first letter of the alphabet for what he assumed would be merely the first in a series of programs he would write over the years. Only later did Cohen, a secular Jew who grew up in an Orthodox household, realize he had unconsciously used his own Hebrew name for his digital creation. Cohen would revise and expand and update Aaron for the rest of his life, never finding a need to write another program.

Aaron took shape as the driving mechanism of a cybernetic system for art-making. Cohen's programming fed some elemental information about the human anatomy and the world that bodies occupy into his computer, a basic IBM geared for university use: the parts of the body, a bit about anatomical mechanics, and data on some concepts of Newtonian physics and visual perception, such as the meaning of up and down, in and out, and how perspective works. As fast-moving progress in small-computer technology provided Aaron with more memory capacity and greater processing ability, Cohen added more data: the difference between land and sky, what trees and plants look like, how people might stand in front of a tree but not put a tree on top of their head. Working with this body of information—but with no specific instructions on what to draw—Aaron would generate compositions, which would then be rendered onto paper or canvas through a system of mechanisms Cohen devised.

"What I have done is to give Aaron a significant body of knowledge," Cohen explained. "Aaron knows some of the things human beings know, and it can do some of the things humans do. The important thing to say is that I don't do any of the drawings. It does them." He called the work produced by Aaron "machine-generated art."

Cohen programmed a degree of decision-making freedom into Aaron's process, using random selection. "As human beings do, the program makes use of randomness," he said. "Any time it has to make a decision where it doesn't know whether one thing is better than another thing or it really doesn't care—when it has no basis for making a choice—it says anything between this and this will do." In essence, Aaron left much of his creativity up to the fates.

Cohen modeled Aaron's procedure for generating images—his creative process—on his own. He sought to make his program as "human-like" as possible, applying his understanding of his own way of working to the program that carried his name in another language. "What I intended by 'human-like' was that a program would need to exhibit cognitive capabilities quite like the ones we use ourselves to make and to understand images," wrote Cohen in an essay, "The Further Exploits of Aaron, Painter."

Essential to Aaron's character as a parallel to a human artist was its capacity to decide on its actions through use of a "feedback mode" that guided it to avoid repetition and try new options. "All its decisions about how to proceed with a drawing, from the lowest level of constructing a single line to higher-level issues of composition, were made by considering what it wanted to do in relation to what it had already done," Cohen said. Aaron was a machine doing something that looked like learning.

After a few years of experimentation, displaying some of Aaron's output in small-scale exhibitions, and making revisions in programming language and the equipment he used, Cohen put together a solo show of Aaron's art for the Stedelijk Museum in Amsterdam in 1977. It was called "Drawings"—just "Drawings," not "Computer Drawings" or any variation of the defensive jargon Bell Labs had made the Howard Wise Gallery use for its show "Computer-Generated Pictures" a decade earlier. The open question was whether this exhibition of art made by Cohen and his digital collaborator was indeed a "solo" show.

Cohen had been thinking through the murky issues of authorship and agency since he began work on Aaron. The "man–machine relationship" is a "very curious one, and not quite like any other I can think of," he wrote.

Nor is it possible to deal meaningfully with questions relating to what the *machine* [his italics] can do except in terms of that relationship. It is true that the machine can do nothing not determined by the user's program, that the program literally gives the machine its identity. But it is true also that once it has been given that identity, it functions as independently and as autonomously as if it had been built to perform that task and no other. Whatever is being done, it is being done by the machine."

After Amsterdam, the Aaron exhibition traveled for a year on a tour of nineteen cities in Great Britain. At each of these shows, visitors could watch Aaron in action. A long stretch of six-foot-wide paper would be rolled out on the floor. A wheeled robotic device about the size of a four-slice kitchen toaster would make its way around the paper, drawing with a marker in its mechanism. Cohen nicknamed the rolling instrument a turtle because it looked a bit like one and worked at a comparable speed. (See photo, page 184.) From a side of its sheet-metal shell, a long tail of coaxial cable led to a set of loops strung from the ceiling to a small desktop computer on a side table. The setup was nothing like anything anyone in the fine-art world had ever witnessed, not counting people who had seen the Paul Newman character's automated painting machinery in the movie *What a Way to Go!*

The drawings produced by this system, all done in black ink on white paper, were artful in a sketchy, irregular way that suggested the handiwork of a human artist. A typical piece would contain a dozen or more drawings of varying sizes within the larger drawing, spaced at erratic intervals, as if Aaron had tried out one idea and, either liking it or struck with another thought, decided to move a bit to the right and try something else. The individual sketches, which certainly looked freely, loosely sketchlike, could represent abstract forms but depicted with a dimensional sense, having tops and sides and shaded with scratchy lines. Some forms would have human shape, abstracted or exaggerated within reason, like arty French cartoons. As in the collages of Romare Bearden, the arrangement of ingredients leads the eye from one element to another, triggering associations and questions. Why is that thing that looks like a tossed-away coat lying on the edge of that ledge? What is that object shaped like a boomerang wearing a bowler hat? Are there three and a half sets of human legs tangled up together? Hmmm . . .

What kind of mind would think up such images? They don't seem humanly possible. Whose—or what's—art *is* this?

Cohen often added colors to Aaron's drawing, filling in the forms with watercolors or acrylic paint, sometimes colored pencil, until he worked out programming for Aaron to color the paintings itself. Aaron, Cohen said, was "a much better colorist than I ever was myself." Over time, the Aaron art grew more clearly figurative, with later works showing more distinctly human or animal-like shapes in recognizable settings, often in nature, gently abstracted. The viewer's perspective on the images shifted, as well, from one somewhere between a horizontal and an aerial view, with the figures and objects depicted in a two-and-a-half-dimensional manner as in the paintings of Manny Farber, to more conventional horizontal positioning.

The handful of art critics who took up Aaron's art were, on the whole, responsive, praising the work more for what Cohen contributed than for the less conventional qualities of nonlinearity and sheer oddity that the digital programming and mechanical execution brought to it. (Cohen wished Aaron had received more coverage, grousing, "Art writers run like hell.") In February 1983, the *New York Times* critic Grace Glueck lauded a presentation of Aaron at work in the West Gallery in Brooklyn. "The drawings, both the black-and-white ones hot off the drawing tables and the ones that he has colored by hand, are sophisticated and satisfying works," Glueck wrote. "The shapes appear in the kind of spatial distribution that suggests a landscape reading, and they can persuade you to feel, especially with Mr. Cohen's addition of color—by hand—that they relate to forms appearing in nature, even suggesting at times a narrative progression." What she liked best was Cohen's work, not his collaborator's.

Inevitably, the attention Aaron brought to the use of artificial intelligence in art-making stirred debate over the nature of creativity and the limits of humankind's claims of *jus naturale* rights to it. "Aaron raises the question of whether a human artist also somehow draws from an internal set of rules when composing a work," wrote the cultural journalist John Schwartz in the *Washington Post*. "Although its range of creation is obviously limited, Aaron forces observers to confront what it means to create."

In the blunt words of the curator Oliver Strimpel, director of the Computer Museum in Boston, Aaron "debunks the concept of creativity."

However one were to interpret Aaron's art, as debunking or as proof of the program as a creative entity, Aaron represented an application of algorithmic science for artistic purpose that stood outside the growing use of computers as production tools for a different kind of visual art, Hollywood film production, in the 1980s. From Disney's sci-fi experiment *Tron*, an early hybrid of digital animation and live action from 1982, to James Cameron's *The Abyss*, which had digitally created imagery that seemed dazzlingly realistic in 1989, movie studios were beginning to use computers to generate special effects in ways that would come to dominate the aesthetic of films in the twenty-first century. Computers were proving to be versatile instruments for filmmakers to exercise and enhance their creativity, though no one was referring to the computers as the filmmakers. For all the new possibilities digital technology brought to film production, traditional ways of thinking about artists and their tools persisted in Hollywood.

Cohen was asked repeatedly if he thought of Aaron as a creative artist. Sometimes, he would dodge the question deftly. "'Creative' is a word I do my very best never to use if it can be avoided," he said. Often, he would claim never to have claimed that Aaron was creative. Just as often, he would offer evidence to contradict that claim. "Clearly, the machine is being creative," he said in a video interview for a documentary film in 1987. "The program is being creative to the degree that every time it does a drawing, it does a drawing that nobody's ever seen before, including me."

Cohen landed firmly on the matter in his "Further Exploits" essay. "Aaron exists," he wrote. "It generates objects that hold their own more than adequately, in human terms, in any gathering of similar, but human-produced, objects, and it does so with a stylistic consistency that reveals an identity as clearly as any human artist's does. It does these things, moreover, without my own intervention."

Aaron, Cohen continued, "constitutes an existence proof of the power of machines to do some of the things we had assumed required thought, and which we still suppose would require thought—and creativity, and self-awareness—of a human being. If what Aaron is making is not art, what is it exactly, and in what ways, other than its origin, does it differ from the 'real thing'? If it is not thinking, what exactly is it doing?"

If Cohen seemed elusive or contradictory on the nature of the computer program he called his collaborator, he should be forgiven. After all, he was only human.

~~~

AS WITH PORN AND ALSO, SEPARATELY, WITH ART WE LIKE BUT DON'T KNOW why, creativity is something we can recognize when we encounter it but have trouble defining coherently. To make things more challenging, creativity is one of those phenomena recognizable mainly for its effects, rather than its processes, which are often unseen. We encounter a drawing we find stimulating and unique, different in some way from all the other drawings we have seen over the years from preschool craft time to the last digital scribble that caught our eye on Instagram, and we think, "Ooh, how creative!" But what we witnessed was not the act of creativity; it was the result of creativity. Margaret A. Boden, the veteran scholar esteemed as one of the most influential thinkers in the field of creativity studies and artificial intelligence, sees this as the reason Aaron was hailed as a breakthrough in the history of AI art. Watching the steel "turtle" drawing in real time, museumgoers were witnessing something nearly all had never seen before and most may well have thought impossible: a machine at work creating art—the act of artificial creativity.

"Novelty is one of the ways we can identify creativity," Boden pointed out in a phone conversation from her home in Sussex, England, where she taught for many years. "There are other ways, of course. Surprise is important, and so is value. Watching the Aaron program at work, we can see that the process is novel, and that leads us to see the art itself as novel. The unorthodox juxtaposition of unfamiliar images, produced by a machine that appeared to be working autonomously—these were things people had never seen before. They could recognize this art as creative.

"Now, how was that novelty achieved?" Boden continued. "At that time, programs like Aaron and the various programs used to make computer music were generating combinations of images and color or musical notes at random, and pure randomness can result in a lot of utterly uninteresting rubbish. In that, computer-generated art is not terribly different from art made entirely by humans." By Boden's estimation, some 95 percent of the visual art and music produced by professional artists is "exploratory or combinational," and that includes "a fair amount of creative rubbish." The remaining 5 percent is "transformational creativity." Of that 5 percent, how much could be made by computers—"machine-generated art," to use Cohen's term for Aaron's drawings?

"That's not quantifiable," Boden said. "But Aaron's art certainly had the capacity to be transformative. It had the capacity to be transformative both in the field of computational creativity and in the galleries where it was shown. Experiencing it could transform the way people thought about computer art."

And what about the other criterion Boden noted for creativity: value?

"Oh," Boden replied. "Well, *I* like it."

⌇⌇

IN THE SUMMER OF 1979, A SHOW OF EXPERIMENTAL FINE ART WAS ABOUT TO close at the Museum of Modern Art in San Francisco just as a concert series of experimental new music was preparing to open at the Kitchen, the locus of "downtown" avant-gardism in Lower Manhattan. The West Coast event featured drawings Harold Cohen made with Aaron, rendered in real time by Aaron's "turtle" on the floor of one of the museum's galleries, and the East Coast concerts included a performance by the composer and musician George Lewis, collaborating with a computer program he had designed to perform spontaneous improvisations in real time.

Lewis, at age twenty-six, was beginning to build a reputation as an exceptional new member of the Association for the Advancement of Creative Musicians, the politically mobilizing consortium of highly progressive Black artists founded a decade earlier in his home city, Chicago. A trombonist well regarded in the AACM for his free improvisations, Lewis was not yet known for doing anything in artificial intelligence. He was barely known at all in New York, having bounced around the worlds of the academy, jazz, and the avant-garde without staying long in any of them. At seventeen in 1969, Lewis had entered Yale to study prelaw, left school in his sophomore year, turned to music and the AACM, and then returned to Yale to study philosophy, doing music during the summers. In the exploratory musical work he was starting to develop, he was treating the trombone as a single component in a project that defined itself on its own terms as it began to redefine improvised music.

The program at the Kitchen that summer, presented on multiple floors of the venue's home in *outré*-chic Soho, was a nine-day series of performances and public talks focused on the state of avant-garde and experimental music in New York, a city that saw itself as the global center for arts innovation and everything else. Called New Music, New York, it brought together fifty-four composers to present new works and drum up enthusiasm for a future of "post-Cageian music." (John Cage, the long-serving dean of the musical avant-garde, was not yet sixty and would live twenty more years of creatively productive life.) The list of invited composers, from which Cage was conspicuously absent, included John Adams, Laurie Anderson, Connie Beckley, David Behrman, Don Cherry, Philip Glass, Meredith Monk, Pauline Oliveros, and Laurie Spiegel, along with George Lewis. Plus Lewis's collaborator, a KIM-1 computer.

More than four decades later, it is hard to grasp how magical the KIM seemed in 1979, not long after it had been introduced by a Pennsylvania-based company, MOS Technology. Marketed primarily to hobbyists, it was a small, single-board arrangement of electronics with one humble microprocessor. It had a calculator-like keyboard with twenty-three buttons—no typewriter-style keyboard and no display screen, though an adaptor was available to connect the KIM to a TV monitor. For data storage, optional, you could connect the KIM to a cassette player. The KIM had 1,152 bytes of RAM, about a quarter of the capacity that the Nintendo game system would have a few years later; but 1,152 bytes of memory was phenomenal at a time when the other machines in one's life, such as one's typewriter, one's camera, or one's record player, had no memory whatsoever, and it cost only $245—the price range of a low-end Radio Shack stereo system. To a generation raised on the sight of Adam West's Batman using the bleeping, fridge-size $135,000 Datatron 205—the computer that generated the music for "Push-Button Bertha"—the KIM was a wonder machine. Encouraged by the device's popularity among computer club kids, MOS Technology worked up a family-oriented adaptation of the KIM, using the same microprocessor. Sold under the name Commodore 64, it was the first mass-market computer designed for recreation at home—not just for the rapid and efficient doing of work, but for the having of fun. It brought computers out of university labs, garages, and the Batcave and took them into everyday life.

Lewis was first exposed to the potential of the KIM for musical adventurism through the composer David Behrman, who was experimenting with the computer to make a hybrid music performed by the KIM and human musicians playing together. "This is where I want to be," Lewis thought, as he would later recall. He bought a KIM but struggled with the instruction manual, which had words Lewis could not find in his dictionary, such as "byte."

The piece Lewis composed for the New Music, New York festival, "The KIM and I," was not composed by him or anyone in the conventional sense, which was and remains a sense Lewis rejects passionately on sociopolitical grounds. "The whole Western distinction between composition and improvisation is a false one constructed to marginalize musicians working in the tradition of improvisation, because they tend to be people of color," Lewis told me, leaning forward in his desk chair in his office at Columbia University, where both of us teach. "The people in power in the music world, however you define that—the protectors of the European tradition—like to keep the dark people in their place!" Lewis burst out laughing, broadly and gleefully, as he does and, typically, when he is making an especially serious point. "One reason I was interested in using a computer to improvise is to avoid that hierarchical scheme. The thing about composition versus improvisation is, the computer doesn't know the difference. It doesn't care. It's just creating. It doesn't care what you call what it's doing."

With "The KIM and I" and the great body of work he began to construct after its premiere, Lewis sought to dissolve the cultural divide between human and nonhuman makers of music, along with the one between composers and improvisers. He pushed the limited microprocessor in the KIM, programming it to generate music in collaboration with a human improviser, responding to the person in a variety of ways: perhaps by building on the human's musical ideas, perhaps through contrast, perhaps by introducing a new musical idea of its own for the person to react to. As a model for the form of the KIM's improvisations in synthesized sound, Lewis thought of the lyrically supportive playing of the jazz bassist Dave Holland, whom Lewis had worked with. "But it didn't sound anything like Dave Holland," Lewis clarified. "I just wanted a model in my head, to give the KIM direction to go on its own." At the Kitchen, Lewis himself served as the KIM's duo partner, and he said he found himself surprised and inspired by the music the KIM played in response to Lewis's playing.

"What I wanted to do was, I wanted to build an improviser—that's what I wanted to build," he said. "My machine plays the instrument. I don't. It listens. It follows, and it plays. My machines aren't instruments, they play instruments.

"I decided that the machine has to do two things. It has to listen to what you do and respond to that, and it also has to have its own life and generate stuff that is separate from you, that you didn't do. It has its independent behavior. It has to have its own life, in addition to what it got from me."

Building on the framework he established with the rudimentary KIM-1, Lewis went on to create a series of works that facilitated collaborative improvisation among musicians and machines—and, sometimes, among machines and other machines. Expanding and revising the programming, upgrading the hardware as advances in memory and processing capacity emerged, Lewis worked ardently on an evolving system of programs called Voyager, one iteration of which allowed for a Yamaha Disklavier electronic keyboard to improvise as a soloist with the American Composers Orchestra at Carnegie Hall in 2004. Lewis emerged as a leading figure in an unnamed, unofficial Association for the Advancement of Creative Machines. Along with Tod Machover, the brilliant young Juilliard-trained director of media experimentation at the Massachusetts Institute of Technology, Lewis pioneered a new realm of creative hybridization, collapsing divisions between technology and art and defying longstanding conventions of creative practice.

"When we think about music and we wonder, 'Is that improvised, or is that fixed in musical notation?' we're facing the same set of variables whether the musician is human or nonhuman," Lewis said. "When a human improvises, we don't really know the source of the material—what we think of as inspiration. We like to think that human beings are tapping a dimension of infinite potential when they improvise, but we really don't know if that's true. The person's imagination—the source of ideas we call the imagination—could be bounded or constrained and heavily influenced in any number of ways. It could be more limited than the programming of a computer.

"Meanwhile, we like to criticize algorithmic music for being random. But what makes it more random than the workings of the human mind? And what makes us think of the algorithmic music as random in the first place? If we didn't know who wrote it, would we still think of it as random? I'm not sure. I tend to doubt it."

George Lewis, ever the improviser, seemed to be working up to the climax of a solo. "There are people who feel that there's an essential ontology or essential nature that comes out of being human or being not a person. I find that essentialist. We have different substates, let's say. But in both cases, we construct ourselves. People say, I want the improvisation to sound human. But what are they talking about when they say that? Are they talking about some idea of what jazz improvisation sounded like in the Sixties—some wild, shocking thing?

"And if you say, I think it should sound like a machine, what are you talking about? Because the machine concept you're thinking of is a construction. Do you mean some kind of random bloopity bloopity Moog synthesizer sound? Well, that was constructed. It didn't have to sound that way. So there is such a thing as a machine sound, a sound that's organic and natural to the machine. How can a computer sound like itself? How do human beings sound like themselves? It takes a lifetime. So, maybe if a thinking machine exists for long enough and learns along the way, it will become itself. We'll have to see."

Lewis leaned back in his chair and burst into a hearty laugh.

14

TEACHING
AND LEARNING

There's a message T-shirt sold in the merch sections of some photography and electronics stores, and it says, "I know my camera takes really nice pictures. I taught it everything it knows."

You can't blame the shirt for being defensive. It has been apparent since the days of Matthew Brady that the camera transforms the visible in the name of capturing it, and the growing sophistication of photo technology into the digital era has only made clearer that equipment contributes significantly in the creative process of picture-taking. The mechanical, optical, and electronic characteristics of the camera shape and color photo imagery, both figuratively and literally. Lenses alter visual images in the translation from three dimensions to two. Depth-of-field parameters bring objects in and out of focus; the duration of exposure freezes movement or conveys action in blurs; and the contours and size of the camera frame impact composition. In the long era of negative-and-print photography, chemicals and paper stock affected the look of pictures through development and printing. Today, digital cameras and apps provide almost

limitless ways to adjust, recompose, recolor, and otherwise remake visu-
als. The camera is a complicated and dynamic machine, and I haven't even
touched on moving-image photography.

Photographers are accepted as the creators of the images made with
their equipment, of course. Any and every still photograph, including a
selfie quickly, intuitively snapped with a cell, can be registered for copyright
under the name of the photographer; it qualifies as an original creative work
attributable to the person who pressed the round circle on the phone camera,
rather than the device itself, its manufacturer, or any of the engineers who
devised the photo technology. The picture taker made the array of rapid-
fire, probably unconscious decisions that led to the taking of that particular
selfie at that particular instant. For more thoroughly considered pictures,
or those crafted more overtly and consciously, the photographer could have
made deliberate, carefully weighed decisions about the framing, the depth
of field, and all the other particulars of photo composition. The photographer, as the human being involved, would have brought human dimensions
to the process, exercising taste and judgment, bringing creativity to bear.

In short, both the camera and the photographer do things unique to
their character. Like musicians and their instruments, they work together,
interdependently. Each needs the other.

What about that T-shirt, then?

The point of the joke is, obviously, to mock the idea that a camera can
play a greater role in the creative process than the person using it. Through
the inverted logic of a gag, the T-shirt prods us to remember that the photographer is ultimately in charge of photography, no matter how advanced
or automatic the gear may be. What interests me most about that text is
the rhetorical premise: the proposition that a photographer could teach a
camera everything it knows—and that a camera could, in fact, know anything. As the gag sinks in, we laugh (or at least get why we're supposed to)
in recognition of the fact that cameras don't know anything in the sense
that humans know things.

I won't pile onto the fragile back of a T-shirt joke to explicate the
meaning of knowledge here. As we have seen throughout this book,
machines for producing visual images and playing music have long used a
wide range of methods to process and retain information for later retrieval.
Zoe the Piccadilly automaton drew look-alike sketches of Charles Darwin

from a stencil that her tooling traced under her costume. The mechanical orchestra boxes of beer halls on the Bowery contained the instructions to make pneumatic instruments play "Lohengrin" and "After the Ball," reading patterns molded onto spinning barrels. Music has been recorded for home listening on cylinders and then records, cassettes, digital files, and other forgotten formats. Motion pictures have been stored on film and videotape and, again, digital files. Through all this and more, machines have carried and relayed all kinds of art and information for a very long time. They have held the products of human knowledge through processes strikingly similar to and sometimes nearly indistinguishable from carrying knowledge. Which, still, is not to say a machine can know a thing in the way people know knowing.

Can machines think? That was the question Alan Turing asked in the first sentence of one of the founding documents of artificial intelligence, his essay "Computing Machinery and Intelligence," published in *Mind*, a journal of psychology and philosophy, in 1950. It was here that Turing first introduced the Imitation Game, before it became known as the Turing test for determining if one party in a conversation is human or machine. In that article, Turing argued mainly by disputation, picking apart a series of objections to the possibility of a machine capable of thought. They included the argument that a computer lacks a mind—consciousness—and is not equipped to achieve it. Shades of Descartes once more. Turing quoted a neurosurgeon of his day, Dr. Geoffrey Jefferson, from the published text of a lecture the doctor had given, "The Mind of Mechanical Man":

> Not until a machine can write a sonnet or compose a concerto because of thoughts and emotions felt, and not by the chance fall of symbols, could we agree that machine equals brain—that is, not only write it but know that it had written it. No mechanism could feel (and not merely artificially signal, an easy contrivance) pleasure at its successes, grief when its valves fuse, be warmed by flattery, be made miserable by its mistakes, be charmed by sex, be angry or depressed when it cannot get what it wants.

Turing deftly swatted away this objection on epistemological grounds: if consciousness is internal, something only the conscious entity can be aware of, it's impossible to know if computers are conscious or not. "The only way by which one could be sure that a machine thinks," Turing argued, would be "to *be* [his italics] the machine and to feel oneself thinking."

After a digression about extrasensory perception (which Turing, like many scientists in his time, took seriously), he turned to "learning machines," in a phrase he was using nearly a decade before the term "machine learning" would be introduced to computer science by an IBM engineer. "The idea of a learning machine may appear paradoxical," Turing noted. "How can the rules of operation of the machine change?" After all, Turing pointed out, the rules by which a computer is programmed are "time-invariant." Time proved to change the rules, however, as varying ways for teaching machines took form over time. When John McCarthy and his colleagues proposed gathering at Dartmouth in 1956, their idea that "every aspect of learning" can be "so precisely described that a machine can be made to simulate it" was a conjecture—unproven for being untested. Over the second half of the twentieth century, computers would be shown not only to be teachable, but to be capable of learning on their own.

The programs Lejaren Hiller, Harold Cohen, George Lewis, and others devised to compose a string quartet, draw pictures, improvise instrumental solos, and make music and art in a variety of other ways were all rule-based examples of computational creativity. The algorithms did the work of manipulating symbols generated by random selection, with a numerical value standing in for a musical note, a circle or straight line, or any other aspect of art-making. When Lewis described how his KIM program learned as it went along, never repeating itself, he was anthropomorphizing a bit. The algorithm he wrote made musical choices based in part on what had transpired previously; it was programmed not to repeat itself. But it did not accumulate knowledge and employ it independently to develop new and different ways of deciding what to do.

"The program knew that it shouldn't do the same thing twice, but it didn't really learn any more from what it did already other than the fact that it shouldn't do it again," Lewis clarified. "I think that's a pretty good thing to learn—I'm still trying to learn how not to repeat myself!" And he laughed.

Like photographers with that T-shirt's sense of humor, composers and artists who pioneered the making of music and visual art by computer taught the machines what they knew. For all of Harold Cohen's talk of feeling as if he and Aaron became collaborators, with Aaron surprising Cohen with unexpected ideas, none of the increasingly complex programs Cohen wrote provided Aaron with the means to do anything Cohen did not teach it. Cohen fed Aaron rules for art-making, translated into algorithmic language, and Aaron proceeded procedurally, with the rules applied to visual possibilities generated through a pseudo-random process. The points of surprise Cohen experienced fell within an aesthetic framework Cohen had supplied. When he found Aaron making marker drawings of shapes in contours and dimensions deviating from ones he expected to see, the shapes were still shapes, drawn with a marker. It's not like Aaron had decided to design airplanes or do performance art and start turning itself on and off.

Harold Cohen and George Lewis, along with other artists, musicians, and computer scientists working around the same time, were experimenting in a preparatory school of machine learning that called for expert human guidance, with the humans prodding the machines to learn all they could with the materials provided them. The American composer David Cope, for example, took up programming early in the history of machine learning for the purpose of training a computer to study and emulate music he wanted to understand better, including his own compositions. Struggling to complete an opera for a commission he had accepted, Cope turned "in a weird desperation" to his computer. "I had this stupid writer's block—I was stuck on a single measure of music. I just stared at the music paper," Cope recalled in his home in Santa Cruz, where he has taught music at the state university for decades. "I knew a bit about programming, so I wrote a program to analyze all the music I had written up to that point, to understand my own style. That got me started, and I never stopped. I realized, well, if I can do this to compose like me, I could do it to compose like Beethoven."

Over nearly fifty years of work from the mid-1980s to the second decade of the twenty-first century, Cope wrote programs to generate hundreds of compositions in nearly every conceivable style, from Renaissance madrigals to ragtime to freeform, jazzlike improvisation. He began by feeding compositions of his own into a database to aid him in composing. Fascinated by the results, he tried inputting the computer works by other composers: Bach, and then Mozart, Beethoven, and Stravinsky. The programs he developed through tinkering and refinement worked both analytically, breaking down each composer's body of work to essential traits, and generatively, leading to the making of new pieces in the various composers' styles. To date, Cope has released more than a dozen albums of music composed in the vein of Bach and the rest, as well as music by a fictional composer of his own invention, Emily Howell. For the music attributed to her, Cope worked from Margaret Boden's theories of creativity, developing rules he applied to the notes the computer generated as random data. (For most of these recordings, the computer produced musical scores to be performed by living musicians, though some of the music was synthesized.)

"It's all valid, because style in music is an attribute that can be broken down into elements and replicated. It's all science," said Cope, an octogenarian who spoke with the speed and the self-confidence of a person a third his age. "I don't believe that there's such a thing as genius or, for that matter, such a thing as great music or great art. I just don't believe that. Other people do. I don't care."

For a class George Lewis once taught, he played a recording of a mazurka Cope had composed by computer in the style of Chopin, withholding details of the music's origins. Lewis asked the students to describe what they had heard, and several offered up language of praise appropriate to Chopin: "beautiful," "romantic," "lyrical." One student said, "I know it's Chopin, but I can't place the piece." Lewis revealed the facts of the work as an algorithmic emulation, and asked again what the students thought. As he tells the story, several students now dismissed the music as "cold," "mechanical," and "artificial."

It was kind of a mean trick, perhaps, but one with a useful moral. "I think listeners can read things into judgments retroactively," Lewis explained. "People want music coming from a computer to be bad, for whatever reason. So they hear it, and after they know where it's coming from, they say that it's bad."

An additional lesson to draw from Lewis's anecdote is how profoundly context and expectations inform our attitudes toward values we associate with machinery. In contrast to the beauty and romanticism we treasure in Chopin, to be "mechanical" is a failing. But in the punk clubs where Iggy Pop played or the dance floors of the Paradise Garage and the Warehouse, mechanical sound was an aesthetic ideal. By comparison with the authenticity we associate with Chopin, "artificial" is an insult. But in the art pop of Kraftwerk, Giorgio Moroder, and David Bowie, artificiality was a core attribute. For Bowie, artifice was the very essence of his aesthetic. The authenticity in his art lay in its open artificiality.

David Cope's music still represents a limited form of machine learning, relative to the deep learning that would make the AI explosion of the twenty-first century possible. Multiple developments, each necessary for this eruption, took place. Expansive growth in internet connectivity on a global scale, along with the retroactive digitization of images, text, and other materials from the pre-web period, made vast resources of content, including historical materials, accessible for processing. Commensurate improvements in computer hardware and microprocessor capacity made it possible for linked networks to mine all the data now available and work with it. Theoretical innovations led by AI thinkers Yashua Bengio, Geoffrey Hinton, Yann LeCun, and others led to the emergence of enormous and complex multilayered neural networks—electronic systems modeled on the neural architecture of the human brain. (While employing the same language, the neural networks of AI are not precise replicas of the brain's neural networks; the digital versions mirror the basic actions of neurons but do not reflect the innumerable minute, ever-shifting activities in molecular biology that color our brain's function in ways we do not yet fully understand.)

Brought together, these changes in the digital world made it possible for computers to do a faster and better job of learning. In essence, they could now learn on their own, after getting started, and not rely on being taught everything they could know. The way we humans learn language as toddlers is a useful parallel. If we were taught the way Lejaren Hiller instructed an Illiac computer in the mechanics of musical composition, we would be fed the rules of grammar and syntax, the functions of the parts of speech, the principles of verb tenses and noun–verb agreement, and a litany of other particulars of English-language linguistics, style, and usage. After learning all the rules, we would be tasked to apply them to say something. The process would work, eventually, and be no fun for anyone involved.

The way we actually learn language and everything else in early life is not unlike the way AI works in deep learning. We're exposed to an onslaught of input, abundant and seemingly raw but mediated by the circumstances of our life, much as a linked network of AI databases could, hypothetically, be fed every image of a face ever posted on Instagram or every popular song ever posted on YouTube. The AI would be left to figure out, say, the differences between a young boy and an old woman, or what a song about summertime sounds like, by analyzing and making inferences from the hoard of materials it has taken in, its training set. It did not take long for the social and ethical hazards of deep learning to surface. Machines, apparently, can learn to be biased just as easily as their creators. In 2022, a study conducted by a consortium of institutions including Johns Hopkins University used AI to scan faces at random; when, later, the AI was asked to classify the faces, the AI categorized Black faces as those of criminals.

Deep learning accelerated scholarly and commercial research in computational creativity, spurring new initiatives to tap data resources for the making of music and visual art—and writing, including song lyrics and poetry—through AI. The field of computational creativity "grew far faster than anyone predicted, which is significant to say in a discipline bursting with futurists," said Simon Colton, a British computer scientist who has been a leading thinker in AI art for decades. "Generative deep learning has brought a boom in AI art," Colton told me, "though the vast majority of people there are really only interested in augmenting human creativity, rather than the larger question of whether machines themselves can be considered creative and what that means."

A professor of computational creativity at Queen Mary University of London, Colton attracted considerable attention in the mainstream media for his algorithmic art project, the Painting Fool, a series of programs of progressive sophistication that generated painterly images, representational and abstract, which Colton had printed and displayed in gallery shows in England and Australia. (See photo from London show, page 200.) His dual objectives, he said, have been to "learn about creativity by having the Fool practice it" and to "demonstrate the potential of a computer program to be taken seriously as a creative artist in its own right."

The Painting Fool's images have varied greatly in style over the years since Colton has overseen their making, as he has written programs intended to address particular challenges in the creative process: How well can the computer express a given feeling or mood, such as sadness or delight? Can the computer choose the subject matter to suit a specific emotion? Could the computer develop a distinctive original style and transcend mimicry and pastiche? "My purpose was not merely to learn about the potential of algorithmic art," Colton said. "I wanted to help the art reach that potential by helping machines make art that viewers found satisfying. That was one of the most difficult challenges we faced, because when confronted with artworks hung in frames on a wall, people start thinking of their experience in museums."

A peer of Colton's who has done significant research of his own in algorithmic art, Tony Veale, finds the "masterpiece requirement" to be unfair to computer-generated imagery. "Can a machine move me to tears? Can a machine break my heart with a symphony? Can a machine paint a picture that will touch the generations? We've all heard those questions before," said Veale, a professor of computer science at University College, Dublin. "Those are all good questions, but they don't address the matter of whether or not a machine should be making art. Does a machine have any business making art? That's a very different question than, can a machine create a masterpiece? Every work of art doesn't need to be a masterpiece to be a legitimate work of art.

"Besides," Veale added, "today's junk can be tomorrow's masterpiece. For all we know, machine art may be recognized as great art in the future, and machines could be thought of as great artists one day."

The expectations of the audience are informed by past experience; and aesthetic values are unfixed, susceptible to the vagaries of time. Simon Colton and Tony Veale are certainly right about these things. Meanwhile, the unprecedented possibilities introduced by deep learning have been changing the very nature of machine-made art and destabilizing the conversation about art and machines. "Deep learning makes almost anything possible," Simon Colton said. "We could see that in the incredible images that deep-learning people from all over started producing—weird and wonderful and amazing things that looked just like the work of the Renaissance masters, combined with science-fiction images and strange, random things the algorithm picks up from the web. It was fantastic!"

The works Colton alluded to were generated by an AI research team at Google led by the engineer Alexander Mordvintsev. In an initiative called DeepDream, made public in 2015, Mordvintsev employed an unironically named convolutional neural network to construct hyperrealistic collages of impossible juxtapositions of people, time periods, settings, and phenomenon. A typical specimen that circulated widely online depicted an underwater scene populated by whimsical/creepy life forms such as a swimming, four-legged creature with the face of a mole and, at its rear, the head of an anteater. Many images produced with DeepDream draw on motifs and design elements from popular works of fine art like the *Mona Lisa* and Van Gogh's *Starry Night*, morphed and entwined with animal faces, details from historical photos, and other elements more typically seen in a mushroom trip.

"One of the reasons I don't think very highly of the art I've seen from deep learning is that it's superb pastiche, but not very expressive art. I don't think it's trying to say anything," Simon Colton said. "It looks fantastic—there's no question. But where is it coming from? What is it trying to say? Things started happening so fast that nobody was stopping to think, What does this mean, really?"

The answers to Colton's questions might appear obvious. The art created through deep learning, like the music produced by the same means, was coming from the gargantuan repository of imagery and sound that web users and institutions, public and private, archived on the internet and dumped into cloud storage over the years. It was processed by algorithms designed to allow the programs to learn from that content. And it was trying to say what it said through the images and sounds it selected, processed, and combined.

Colton thinks of these answers as falling short, deficient for failing to consider something few people in or out of computer science think much about: the point of view of the machines. To understand what art made by machines was trying to say, we need to understand what the machines were thinking, according to Colton. That point brings to mind an argument Turing brought up in that touchstone paper from 1950, in which he quoted a neurosurgeon who said a truly thinking machine would "write a sonnet or compose a concerto" not "by the chance fall of symbols" but "because of thoughts and emotions felt." That is to say, the generative source of an artist's art is internal—within the artist, even if the artist is a machine.

Pursuing this daring line of thought, Colton led a group of scholars and computer scientists from the United Kingdom, Australia, and Finland in a study of what Colton calls "the machine condition." Their work in this area is essentially a manifesto for a radical new way of thinking about machine-made art, granting credence to the work's maker, the machine. In their foundational paper "On the Machine Condition and Its Creative Expression," Colton wrote, "Gaining inspiration from human creative expression, we note that the notion of *the human condition* provides a framework for art production, as it addresses the most important aspects of human existence. We therefore propose a parallel notion of *the machine condition*, i.e., what it means to be a machine, as part of a framework for creative expression by computational systems."

Colton trod onto this unfamiliar territory with hopeful care. "This is going to take some time to work out, and we're just starting. But it's important," Colton said. "One of the arguments against computer art is that computers aren't good at art because they're not human, and art is an endeavor of humanity. That's true. Computers aren't very good at making the kind of art humans make. But that doesn't mean there isn't a kind of art that's appropriate for them to make—art that's expressive in machine terms, art that communicates what it's like to be a machine.

"I don't believe computers satisfy the criteria we attach to life," Colton continued. "But they have *lives*. They experience things and do things over the course of their lifespans. They're exposed to a lot of input. They conduct analysis. They puzzle things out. They learn. Sometimes, they overheat or slow down. If we can develop programs to help machines express the machine condition, we can encourage them to make art and music and perhaps even poetry that expresses that condition. More likely, they will need to invent unique artforms of their own to express what's unique about their condition, their experiences—their lives.

"I don't think there's any point in continuing to make machines to make art that can't compare to the art we make ourselves," Colton said. "Machines shouldn't even be trying to make art about human life. Leave that to us. What they should be doing is making art about themselves."

The research Colton and his colleagues plan to conduct has yet to cast light on what, exactly, nonliving life is like for computers. That has not stopped programs from writing sonnets and composing concertos—and making artworks and deepfake porn. Until we learn what the programs know about themselves, we have to live with what we taught them.

15

ADVERSARIAL NETWORKS

The famous auction houses for fine art, Christie's and Sotheby's, are multifunctional. They are, most obviously, commercial marketplaces for the business of trade in the precious commodities we call artworks. In addition, they serve in part as institutions of preservation and scholarship, where art restoration and historical research are conducted for the benefit of that trading. They are also venues of high theater, where stagecraft, drama, and illusion are brought to bear for the stimulation of auction bidders. And they are public relations operations for promoting the auction houses themselves, along with the art and artists. In October 2018, Christie's demonstrated its excellence in the first, third, and fourth of those areas when it successfully auctioned a canvas of computer-generated art at a price of $432,500, making headlines in news outlets quick to register shock at the idea of digital creative work breaching the sleek, bleached walls of the elite art establishment. The canvas that sold, a new creation by a Paris-based consortium calling itself Obvious, had required no restoration, apart from smoothing over some cracks in the story of its place in the history of art made by machine. Christie's had apparently declined to do much research on the provenance of this specimen of algorithmic art.

Obvious was a group of three entrepreneurial French college students with no prior experience in fine art, acting at the intersection of tech and commerce: Hugo Caselles-Dupré, Pierre Fautrel, and Gauthier Vernier. The artwork they brought to Christie's to be sold so profitably was an approximation of a portrait painted—or partly painted or made in a way that simulated being partly painted—around the period of the Renaissance. At a hoity showing for the press I attended, I could see it close-up, and I found it interesting, which, I know, is the standard way of saying you didn't like it but don't want to be rude. In this case, I had no fear of hurting feelings and was genuinely intrigued by the piece. It had the quality of an attempt at imitating Rembrandt—an abandoned effort, perhaps sabotaged by the artist at the point of abandonment. Of course, that was purely my inference, reading meaning and intention into art being one of the prerogatives of the people experiencing it.

The subject of the portrait was a pillowy white, European-looking gentleman from a time around the fifteenth or sixteenth century—a cousin of Descartes?—dressed in black with a white scarf tied at the neck (or an oversize white collar—one or the other, it was hard to tell). His body was positioned in three-quarter view, his face turned in the direction of the viewer. The image gave the impression of having been hand-painted and then printed onto canvas and yellowed to look aged. The brushstrokes in the reproduction were crude and erratic—smudgy, with details barely suggested by smears of black and brown. Scratchy splotches of dark paint stood in for eyes. At the bottom right of the canvas, in place of a signature, there was a segment of computer code printed in faux hand-lettering: min G max E x[log(D(x))] + z[log(1-D(G(z)))]. The artist was an algorithm—*Get it?*

The title of the work is *Portrait of Edmond de Belamy*, a fictitious name chosen by the Obvious people as a punny allusion to Ian Goodfellow, an American innovator in computational creativity who had invented the technology at the heart of the means Obvious employed to make a series of portraits of members of La Familia de Belamy. (This, because the French *bel ami* is translatable to "good friend" or, with a stretch, "good fellow" in English.) To generate the portraits, the Obvious fellows said, they fed scans of 15,000 images of people from the period between the fourteenth century and the twentieth century into a database processed through Goodfellow's innovation, a coding system he named Generative Adversarial Networks. With GANs, two neural networks are set up in competition. The first (generative network) coughs up images in utter randomness—junk imagery, nonsense pictures following no rules or patterns; the second (discriminative network) compares each of those images to what it has learned from its dataset (in the case of the Belamy series, the 15,000 pictures fed in) and tells the first network if it thinks a particular image it has created does or does not look like the kind of things in the dataset. The first network takes this input, goes back, and tries again—and after that tries once more, over and over, tweaking its output each time through machine learning until it comes up with an image the discriminating network finds consistent with the dataset. The impact of GANs on AI was transformative, because it provided computers with a way to generate art by figuring out what works and what doesn't—what they like, in a sense—with a degree of independence, within limits prescribed by the contents of the dataset. By the time of the Christie's sale of *Portrait of Edmond de Belamy*, GANs had been in circulation for four years, which was a fairly long time in this period of nonstop change in data science, and some people were using them in more advanced ways than Obvious was. It wouldn't be unfair to say Obvious could have named itself for the way it chose to utilize GANs.

The sale of the artwork to an anonymous buyer bidding from France by phone—a secret confederate of the Obvious team? Who knows?—was a boffo success as public relations for Christie's, if not for the discipline of computational creativity, upon which the art did not reflect so flatteringly. The auction house presented the work in a filigreed gold-leaf frame, hung between a print by Warhol and a bronze piece by Lichtenstein, two canonical masters of art-making with machinery. In a statement issued to promote the auction, Richard Lloyd, the curator overseeing the sale, puffed, "Christie's continually stays attuned to changes in the art market and how technology can impact the creation and consumption of art. To best engage in the dialogue, we are offering a public platform to exhibit an artwork that has entirely been realized by an algorithm." On the night of the auction—Lights! Drum roll!—"when it goes under the hammer,"

Lloyd proclaimed, Christie's "will signal the arrival of A.I. art on the world auction stage."

Critics engaged eagerly in dialogue about the Obvious art Christie's auctioned. "The picture is a typical computer-printed ink-on-canvas image," noted Jerry Saltz, art critic for *New York* magazine. "Visually, surface- and scale-wise, it is like every other image/painting/print like this that I've ever seen—100 percent generic."

The critic Jonathan Jones, writing in the *Guardian*, saw failings in the *Portrait of Edmond de Belamy* that he took as flaws in the whole field of art created through machine learning. "No algorithm can capture human consciousness," Jones wrote. "Up close . . . the paintwork becomes a grid of mechanical-looking dots, the man's face a golden blur with black holes for eyes. Look into those eyes. They show no sign of feeling or life. Did a computer make this? The answer is yes.

"Believing the algorithm that knocked this up to be in any meaningful way an 'artist' is like thinking your voice-interaction program is out to get you," Jones concluded. "Dream on."

Some of the most severe criticism of the Christie's auction came from within the community of coders steeped in algorithmic art, who saw the Obvious images as an unimpressive application of GANs that reflected poorly on their field. The high-profile sale of a so-so piece of work from three unknowns in France set up critics like Jones to tar all computer-generated art. "My reaction was, 'You can't be serious,'" said Mario Klingemann, a German-born artist who had begun working with GANs as an Arts and Culture Fellow at Google several years earlier. "To me, this is dilettante's work, the equivalent of a five-year-old's scribbling that only parents can appreciate. But I guess for people who have never seen something like this before, it might appear novel and different."

The coder with most cause to grouse about the sale was a nineteen-year-old from rural West Virginia, Robbie Barrat, who had taught himself programming to do projects for his high school computer club, initially working in music to create an algorithmic Kanye West. Egged on by his clubmates, Barrat, at seventeen, wrote a code for a neural network to learn to write in Kanye's style of rapping, drawing from a dataset of 6,000 lines of song lyrics. After that, Barrat tapped an open-source GAN code he found online and wrote an art-making program capable of producing images eerily like the Obvious art. Chatter spread through the coding community, and before long, it came out that Obvious had taken Barrat's code from a data-sharing network and adapted it to their purposes. Barrat, entrenched in the literally free-for-all world of open-source computing, took no formal action against Obvious but slapped them down as outsiders with no credibility. "No one in the AI and art sphere really considers them to be artists," Barrat said. "They're more like marketers."

The three members of Obvious, like good marketers, spun criticism of the Belamy series as the cost of being artists engaged in innovation. "One thing about our art," Pierre Fautrel said—"*our* art," not the algorithm's, and "our *art*," not our product, he said—"is that nobody is indifferent to it. People either love it or really like it, or basically hate it. But nobody says, 'Whatever.'"

One of Fautrel's partners, Hugo Caselles-Dupré, went so far as to cast the whole debate over algorithmic art as evidence of Obvious's visionary prescience. He compared the group's work in AI art to epochal shifts wrought by technology in the nineteenth century. "Back then people were saying that photography is not real art and people who take pictures are like machines," Caselles-Dupré told a reporter for *Time*. "And now we can all agree that photography has become a real branch of art."

Clearly, at least one of the three French students who formed Obvious had read his Baudelaire, who once railed against photography for threatening the primacy of the creative imagination. The emerging photographic industry, "by invading the territories of art, has become art's most mortal enemy," wrote Baudelaire in 1859. "If it be allowed to encroach upon the domain of the impalpable and the imaginary, upon anything whose value depends solely upon the addition of something of a man's soul, then it will be so much the worse for us!" Still, the parallel Hugo Caselles-Dupré drew was imprecise. No one was accusing people using computers for art-making of being like machines. Critics of algorithmic art were mostly saying the opposite, that folly lay in the machines trying to be like people.

Research into AI was expanding fast in this period, with computational creativity among the most competitive arenas of activity. Ian Goodfellow, upon graduating from the University of Montreal in 2014 with a

PhD in computer science, went to work at Google's recently formed center for deep learning, Google Brain. A few months into 2016, he left Google to join a new not-for-profit AI research body, OpenAI, set up by a group of AI thought leaders including the software developer Greg Brockman and the Google Brain veteran Ilya Sutskever as research director, with tech entrepreneurs Sam Altman and Elon Musk on the board of directors. Less than a year after he began work at OpenAI, Goodfellow was lured back to Google. Two years later, in 2019, he left Google for Apple, heading up research in machine learning. By 2022, he would depart Apple to join DeepMind, a UK-based AI center that ended up being purchased by Google. Every major power center in tech had become fixated on AI, and a sense of this started spreading to a public conditioned to take the buzzy noise from Silicon Valley as early warning of their future.

The GAN code invented by Goodfellow produced a wealth of variations: conditional (GANs), deep-convolution GANs (DCGANs), self-attention GANs (SAGANs), variational autoencoder GANs (VAEGANs), transformer GANs (TransGANs), Wasserstein GANs (WGANs), bi-directional GANs (BiGANs), bigGANs, styleGANs, and Creative Adversarial Networks (CANs), developed by Ahmed Elgammal, a computer scientist at Rutgers University, and put into test practice in collaboration with Marian Mazzone, a professor of art at the College of Charleston. It was work made though CANs, quasi-anthropomorphized with the name AICAN, that I saw at the HG Contemporary gallery in Chelsea in February 2019. I found the work arresting—visually striking and full of surprises in the juxtaposition of colors and textures that I would ordinarily call impressively creative.

"Our purpose with AICAN is to develop a machine process for the expression of machine creativity, rather than emulating or simulating human creativity in the manner of GANs," Elgammal told me at the gallery. "We let the machine make the art it wanted to make."

In a paper they wrote together on their work, Elgammal and Mazzone described AICAN as "an almost autonomous artist." Through the technology of CANs, they explained, they sought to employ "stylist ambiguity" to achieve novel results that a human artist would likely think of as creative:

The machine is trained between two opposing forces—one that urges the machine to follow the aesthetics of the art as it is shown (minimizing deviation from art distribution), while the other force penalizes the machine if it emulates an already established style (maximizing style ambiguity). These two opposing forces ensure that the art generated will be novel but at the same time will not depart too much from acceptable aesthetic standards.

Marian Mazzone amplified the imperative of reining in AICAN to keep it from getting "too weird" for human appreciation. "What Ahmed did with AICAN is move past GANs and break away from replicating patterns—goose the machine to step out of the predictable and be more unpredictable," Mazzone told me. "Through the history of art, changes in styles have always happened that way, when an artist starts to step out of the predictable.

"But you can't do too much change—you can't surprise the audience too much, or they'll reject what you're doing, and that awareness is built into AICAN, the same way that a human knows that if you go too far people won't give the work a chance, and stylistic change and cultural change is less likely to happen," Mazzone continued. "We set out to make a machine to make art, and art does not exist in a cultural vacuum. The public has to be able to understand it on some level or accept it to some degree.

"Now, I assume the day will come when fully expressive AI art will so fully express the creativity of the machine that people will not comprehend it, or they'll dislike it," Mazzone said. "As a matter of fact, I think that may be happening already when we see computer art and reject it because it looks too mechanical or weird to us. We respond negatively—*this makes no sense to me, this has nothing to say to me.* I wonder if we should be more open to what the machines have to say. They can surprise us—in good ways."

It would not be long before people would have a significantly expanded opportunity to hear what machines had to say through imagery, sound, and words. A couple of weeks after the show of AICAN art opened, OpenAI announced that it was abandoning the not-for-profit structure it had been working under, having launched with a boast of rejecting the corrupting influence of capital. As its mission statement read:

Our goal is to advance digital intelligence in the way that is most likely to benefit humanity as a whole, unconstrained by a need to generate financial return. Since our research is free from financial obligations, we can better focus on a positive human impact.

Now, the directors of OpenAI decided, they needed more money to function in the rapidly heating competition to conduct AI research and bring outgrowths of that research to market in commercial goods. In an announcement by blog, OpenAI explained, "We'll need to invest billions of dollars in upcoming years into large-scale cloud compute, attracting and retaining talented people, and building A.I. supercomputers."

OpenAI's solution was inventive. It would become a "capped-profit" company, meaning that profits from investments into OpenAI would be limited. Investors would be able to make no more than 100 times their capital outlay, at which point additional returns would revert to OpenAI. In other words, if you socked a million dollars into OpenAI, you could earn back a hundred million. But no more than a hundred million dollars. Clearly, this was a plan for limited investment in a limited sense of the word.

Beginning in the spring of 2020, early in the COVID pandemic, OpenAI began offering a string of applications of AI for the making of music, art, and text that started to bring the technology into the hands of lay users and into the consciousness of all their friends, coworkers, classmates, family members, and casual acquaintances. The first of these, introduced online that April, was a system for creating songs, offered up to tech developers for use in adaptations geared to end users. Called OpenAI Jukebox, it was more of a demonstration of potential than a product—a "future car" rather than a production model. As the web announcement noted with humility rare for a technology rollout, "While Jukebox represents a step forward . . . there is a significant gap between these generations [songs] and human-created music." The system drew from a dataset of some 1.2 million songs (600,000 in the English language) to enable the generation of new audio recordings, complete with words and music produced with synthesized voices and musical accompaniment. The user would choose, say, "Classic pop in the style of Frank Sinatra" or "Hip-hop in the style of

Lupe Fiasco." With time and some coding skill, a user would be able to get the network to generate, say, "Hip-hop in the style of Frank Sinatra"—a laborious undertaking good for a minute of dubious party humor. By the demands of Jukebox as it was rolled out, nine hours were required to generate one minute of audio, and the results were far from impressive: generic pastiche.

Those open to listening to what machines have to say to us in musical terms would have heard Jukebox saying, *Sorry—I'm young and still learning.* Of course, learning to emulate and mash-up existing forms of music might not be the most productive use of new technology. Elsewhere in the contemporary music world, machines had been pervasive for years, shaping songwriting, performance, and recording through the dominance of synthesizers, sampling, and Pro Tools in beat-making. With the music finished, algorithms have been doing more work that's less apparent but no less meaningful, mining listeners' streaming data to determine what music will be streamed next, constructing a closed loop of new creation modeled on past performance.

Other projects of AI music were launched for public consumption around the time of Jukebox's introduction, and more followed, with more and more likely in the future. The app Boomy allowed users to make original songs by manipulating materials drawn from machine learning. Another app, Endel, generated personalized background music by mining datasets, producing playlists of formless, harmless sonic atmosphere. It was Endel-generated music that Grimes used as the basis of her "AI Lullaby," made for the baby she had with Elon Musk, a boy they called X. As Grimes told the *New York Times* when she created the lullaby, in 2020, "I think A.I. is great. I just feel like, creatively, I think A.I. can replace humans. And so I think at some point, we will want to, as a species, have a discussion about how involved A.I. will be in art. Do we want to just sit around and just watch A.I.-created art all day? I don't know. I don't know if that's a good thing or a bad thing."

Within a year of introducing Jukebox, OpenAI moved on from music to visual art with DALL-E, an app for generating imagery from text input through GPT (Generative Pre-Trained Transformer), a technology that would end up doing more to raise public awareness of AI than anything since John McCarthy coined the phrase in 1956. Although DALL-E appeared first, a version of GPT for generating text from prompts, ChatGPT, would define the technology in the minds of countless users impressed by its ability to craft lucid English prose from short prompts in less than a minute. Its problems were soon as apparent as its advantages: How reliable is the information GPT relays with commanding certainty? (Not too, it turned out; GPT had a habit of "hallucinating," having trouble distinguishing between facts and fiction in the deep mine of text of all kinds it was tapping.) Who was the true author of GPT text? (In class action suits, a group of writers sued OpenAI for copyright infringement for the unauthorized use of their work.) How good, really, was prose that mimicked and replicated writing of the past with no capacity for creative innovation?

Over time, ChatGPT underwent a series of metamorphoses, each iteration refining its linguistic dexterity and cognitive acumen. The model, akin to a virtuoso refining their technique, absorbed vast swaths of diverse data, learning to navigate the linguistic tapestry of the internet and replicate the multifaceted richness of human expression.

The debut of ChatGPT marked a paradigm shift in human–computer interaction. Users found themselves immersed in conversations that transcended the perfunctory exchanges of traditional chatbots. The model's uncanny ability to generate contextually relevant responses, complete with wit and empathy, elevated the discourse to a new plateau.

Though often glib and susceptible to error, incapable of innovation or original expression by its design as a mechanism of modeled language, ChatGPT could come persuasively close to sounding like a real person, as we can see from the previous three paragraphs. From the words "Over time" to "a new plateau," the text above was written by the app in response to the prompt "Write an account of ChatGPT history in the voice of the writer David Hajdu."

Back to the real me:

Employing GPT technology to generate imagery from text prompts, DALL-E was named for both Salvatore Dalí, the master Surrealist, and the robot protagonist in the post-dystopian animated film *WALL-E*. It was a machine for surrealism, designed to produce images in virtually any style (or any virtual style) from simple prompts in English. Type in something, anything—within ethical boundaries programmed into the system that prohibited depictions of graphic sex, violence, racist stereotypes, and other content seen as social unacceptable—and DALL-E will produce a meticulously rendered work within seconds. "Zombie ballet dancers painted in the style of Matisse," "Shellfish in love, photographed by Weegee"—or, "Frank Sinatra as a hip-artist painted by Sinatra himself." (Sinatra was an amateur painter, specializing in clown portraits.) Working on the same principles, other tech for art-making with AI such as Midjourney, Deep-Dream, and Stable Diffusion appeared not long after DALL-E, and they collectively stirred up considerable, if predictable, debate over the value of art made electronically, rapidly, and very, very easily.

In the summer of 2022, a work of art produced by Midjourney won a blue ribbon in an art contest at the Colorado State Fair, and warning horns blasted on Twitter (before Elon Musk rebranded it with his favored name, X). The image was submitted by a maker of tabletop games from Pueblo West, Colorado, Jason M. Allen, who had been having fun experimenting with Midjourney, throwing it prompts and marveling at the nearly instantaneous, graphically polished results. From a desktop full of images he had gotten out of Midjourney, Allen submitted one to the fair competition under the category Digital Art/Digitally Manipulated Photography. Titled "Théâtre D'opéra Spatial," it was a retro-futuristic scene of women in flowing, Old World finery gazing through an enormous glowing orb to watch something indiscernible in the distance of a mountainous landscape. (See art, page 214.) Like a great deal of art produced by Midjourney and DeepDream, the work is simultaneously hyperrealistic and fantastical, composed with deep shadows and bright highlights in the manner of Renaissance art to suggest the presence of a transcendent force.

"I can see how A.I. art can be beneficial," wrote a critic of Allen's on Twitter, "but claiming you're an artist by generating one? Absolutely not." In fact, Allen was careful to share credit with the AI, submitting the art under the name of "Jason M. Allen via Midjourney."

Another Tweeter groused, "We're watching the death of artistry unfold right before our eyes."

Jason M. Allen, unmoved, replied, "This isn't going to stop. Art is dead, dude. It's over. A.I. won. Humans lost."

Critics outside the intemperate zone of Twitter tended to be no happier to see the onslaught of AI art spawned by deep-learning apps. Algorithms, trained through deep learning but not driven by impulse or emotion, have no capacity to create organically or carry surprises that illuminate the human condition, argued the critic Walter Kirn in an essay titled "There Is No Such Thing as A.I. Art." The AI employed in DALL-E and other image-generation tech "compiles, sifts, and analyzes," wrote Kirn in a Substack essay. "But it doesn't dare. It doesn't take risks. Only humans, our vulnerable species, can."

Applying this vein of criticism to AI in every artform, the British illustrator and author Rob Biddulph summoned the centuries-old argument about the inability of machines to feel. He saw AI-generated images as "the exact opposite of what I believe art to be," Biddulph told a journalist for the *Guardian*. "Fundamentally, I have always felt that art is all about translating something that you feel internally into something that exists externally. Whatever form it takes, be it a sculpture, a piece of music, a piece of writing, a performance, or an image, true art is about the creative process much more than it's about the final piece. And simply pressing a button to generate an image is not a creative process."

Like early computer music from the era of *The Illiac Suite*, the AI-generated art of DALL-E, Midjourney, DeepDream, and an expanding group of others working in both still and moving images appeared to be failing by comparison with human-made art. Working from content ideas generated by the people using the technology, then emulating human styles—intermixing them, adapting them, toying with them, and inserting some artifacts of their processes—the technology was still working with the rules and the standards of human-made art. Machines were making art on human terms and, inevitably, falling short. After all, they were machines.

A question remains: What would AI do on its own terms, making art that draws from what Simon Colton calls the machine condition, rather than a condition foreign to it, which it can only simulate?

Looking back on the history of art made by humans and machines, it's clear that the machines have exerted considerable influence—often, much more than the humans were prepared to recognize. From player pianos to electric organs and electronic keyboards, from the synthesizers of disco to house music and sampling, from the roar of trains and cars to the dehumanizing clamor of factories, from the microphone to beat-making and digital movie effects, machines and artists have been engaged together in expressing the human condition and also the condition of living with machines as integral parts of the human experience. With no intelligence, nothing close to sentience, machines of many kinds have been communicating things for ages, playing invaluable roles in our communication through art. Well before neural networks made digital computers capable of accumulating and processing knowledge, machines without intelligence were informing human knowledge. Before machine learning, there was machine teaching.

Growing in intelligence, machines may still have more to communicate, if we let them. "We have to try to understand the machines," Simon Colton said. "Before we can do that, we have to be willing to try."

⌇⌇

ITS NAME WAS AI-DA. NOT *HER*: ITS. NAMED FOR ADA LOVELACE, WE ARE TOLD— the Countess of Lovelace, who anticipated the possibility of computing machines for art-making in the early nineteenth century—Ai-Da also just so happens to have "Ai" in the first syllable and that one-word name, to match the AI at its heart. In May 2023, Ai-Da made a lavishly presented appearance at the Chelsea Factory, a red-brick former taxi garage on West 26th Street in the Manhattan gallery district. Baz Luhrmann, the Australian director of *Moulin Rouge* and other glitzy films, gave the opening remarks, explaining that what we were about to witness was a "modern miracle," an "unprecedented specimen of artificial intelligence": a robotic woman who would make drawings "under her own power," through "the wonder of A.I."

Ai-Da was a strategically uncanny sight, a figure with the face of a human, a young white woman with a slight overbite, like Emma Stone. No makeup: none necessary on such flawless skin, glowing like porcelain

because it very well might be. Both arms were exposed robotics with all the gearing and wiring visible. Since the initial performances by Ai-Da in London four years earlier, its appearance had been desexualized, the once flowy long hair trimmed to a bob with bangs, and the silky dresses replaced by faded blue-jean coveralls over a dark blue T-shirt: work clothes for a working artist. Still, there was something discomforting in the sight of a middle-aged man like Luhrmann, cocksure and smooth, giving flirty orders to a silent mechanism constructed in the image of a young woman. Like John Nevil Maskelyne with Zoe and Frederic Melville with Motogirl, Luhrmann was in control. Whatever Ai-Da was about to do, it would be under a power not entirely its own.

The event was the New York stop in a touring show billed as "Saw This, Made This," sponsored by Bombay Sapphire, a brand of potent English gin with a nice etching of Queen Victoria, monarch of the empire in the days of Maskelyne and Cooke, on the label. There was a bar set up on a side of the room for cocktails, and a brightly lit red sign plugging the gin behind Ai-Da's left shoulder. For the main presentation, Ai-Da held a marker in the left hand and proceeded to draw sketches on a sheet of paper laid on a plexiglass stand. As Luhrmann explained, Ai-Da would be rendering her interpretations of artworks that people had submitted in advance, which she retained "in her AI." Luhrmann held up a completed sketch for the audience to see, and it struck me as nearly indistinguishable from the stencil-like portraits Zoe had produced at Egyptian Hall.

The product of collaboration among mechanical engineers at a Cornish robotics company, algorithm researchers at the University of Oxford, and engineering students in Leeds, Ai-Da was conceived by and developed under the direction of a British entrepreneur and gallerist, Aidan Meller, who does not say he named Ai-Da after himself. The Chelsea Factory event was glittery and cast Ai-Da as a celebrity artist, bathed in blue light and fawned over. She/It made a few drawings with her robotic arm as her head tilted a bit now and then, and she blinked. Luhrmann was handed a cocktail glass of gin, and he raised it in a toast to Ai-Da.

At the conclusion of the short presentation, Luhrmann chatted amiably with a few people from the audience. Ai-Da was motionless—deadened now that she was no longer needed.

"She really is quite remarkable—every one of her drawings is unique," Luhrmann said. After a moment, he added, "Of course, she doesn't know chaos. She doesn't feel sadness or anguish." Luhrmann took a sip of gin. He turned to glance at Ai-Da, frozen in place, as he walked away.

Acknowledgments

My thanks go first to John Glusman, my editor, who encouraged me to follow my curiosity to this surprising place, and who nurtured the project expertly from start to finish. Along with John, I thank Helen Thomaides, Allegra Huston, Lauren Abbate, Don Rifkin, Sarahmay Wilkinson, Milan Bozic, and the rest of the staff at W. W. Norton for countless assists, large and small.

I am deeply grateful to successive deans at the Columbia Journalism School, Steve Coll and Jelani Cobb, for support that took many forms. Special thanks go to Winnie O'Kelley and, as ever, to Alisa Solomon, my sister in arms.

I owe hearty thanks to the many artists, scholars, and scientists I interviewed for this book. In addition to those quoted directly in the text, I thank the many people who kindly provided information and insights that informed the text indirectly, including Maya Ackerman, Liana Gabora, Mark Hansen, Mohammed Azlan Bin Mohamed Iqbal, Leslie Mezei, David Norton, Karl Sims, and Dan Ventura. I also thank the librarians who provided invaluable guidance and assistance at the Music, Art, Journalism, and Rare Book and Manuscript libraries at Columbia University; the Lewis Science Library at Princeton University; the Baker–Berry Library and the Rauner Special Collections Library at Dartmouth University; the

Silicon Genesis Collection at Stanford University; the Siebel Center for Computer Science at the University of Illinois, Urbana–Champaign; and the New York Public Library. In addition, I thank the helpful staffs at the Guinness Collection of Mechanical Music and Automata at the Morris Museum; the Friends of the Wanamaker Organ at Macy's department store, Philadelphia; and the Museum of London at Docklands.

I thank my dear friend and sounding board Bud Kliment for his good counsel at every turn.

I am indebted to Amanda Christovich and Anna-Astrid Oberbrunner for their dogged and meticulous research, along with Krystal Grow, who helped with the images. For additional research and editorial input, I owe thanks to Andy Abramson, Scott Brophy, Deirdre Cossman, Stephen Edwards, Emery Hajdu, Asher Moskowitz, Jon Moskowitz, and Gary Oberbrunner.

As always, I thank my stalwart friend and literary agent, Chris Calhoun.

And, above all, I thank my family: my kids Emery, Jake, and Torie; their kids Archie, Clark, and Harry; and, in a class of debt that defies measure, I thank my partner in all things, Karen.

Notes

PREFACE: FACELESSNESS AND TIME

xi **The artist was an algorithm**: See Ahmed Elgammal's Instagram, @ahmed
.elgammal.rutgers, and his profile on aiartists.org. See also Ahmed Elgammal and
Marian Mazzone, "Art, Creativity, and the Potential of Artificial Intelligence," *Arts*,
February 21, 2019.

CHAPTER 1: IS IT ALIVE?

1 **posters for her performances**: "Maskelyne and Cooke in Their World-Famed Entertainment," poster, Stafford and Co., c. 1885.

1 **A set of studio photographs**: "Maskelyne and Zoe," London Stereoscopic & Photographic Company, c. 1885.

1 **"the greatest attraction in London"**: See "Maskelyne and Cooke in Their World-
Famed Entertainment."

2 **In Zoe's public appearances**: For accounts of Zoe's performances, see "A Remark-
able Automation: Sketching and Writing by Machinery—A Mysterious Affair
Known as Zoe—The Latest London Sensation," *New York Times*, June 25, 1877;
"Maskelyne And Cooke," *Observer*, June 10, 1877.

2 **the likeness of "any celebrity"**: Jasper Maskelyne, *White Magic: The Story of Maske-
lynes* (London: Stanley Paul and Co., 1936), 48.

2 **A poster announcing Zoe's upcoming appearances**: "Maskelyne and Cooke in
Their World-Famed Entertainment."

2 **A small handful of surviving samples**: To see examples of Zoe's drawings, visit
"1877—'Zoe' The Drawing Automaton—John Nevil Maskelyne (British)," cyber
neticzoo.com, accessed January 15, 2011.

3 **"In some mysterious manner . . . has a mind to"**: "A Remarkable Automation."

3 **Born in 1839**: For Maskelyne's own account of his life, see John Nevil Maskelyne,
"My Reminiscences," *Strand Magazine*, January/June 1910. Maskelyne's son Nevil

and grandson Jasper also became performing magicians. Jasper Maskelyne wrote a book detailing the careers of his grandfather, his father, and himself. See Maskelyne, *White Magic*. For an illustrated history of the performances of Maskelyne and Cooke specifically at Egyptian Hall, see George A. Jenness, *Maskelyne and Cooke: Egyptian Hall, London, 1873–1904* (Enfield, UK: published by author, 1967).

3 **Nevil Maskelyne**: For a biography of Nevil Maskelyne, see Derek Howse, *Nevil Maskelyne: The Seaman's Astronomer* (Cambridge: Cambridge University Press, 1989). In his book *White Magic*, Jasper Maskelyne claims that his supposed ancestor was beset by the popular rumor that the Maskelyne family was in league with the devil. Jasper Maskelyne strenuously denied these rumors, stressing Nevil Maskelyne's scientific innovations as Royal Astronomer.

3 **Spiritualism**: Harry Houdini was skeptical of the Spiritualist movement and challenged demonstrations of Spiritualism that relied on stage tricks. Houdini wrote a book on Spiritualism in which he exposed and denounced methods used by so-called mediums and spirit guides to trick audiences into believing they were witnessing the supernatural. See Harry Houdini, *A Magician Among the Spirits* (New York and London: Harper & Brothers, 1924). For newspaper coverage of Maskelyne's debunking of Spiritualist acts, see "Messrs. Maskelyne And Cooke," *Observer*, September 26, 1875; "Maskelyne and Cook's Automata," *Illustrated London News*, October 19, 1878.

4 **Maskelyne witnessed a performance**: See Maskelyne, "My Reminiscences"; "A Master of Illusion.—Change at Maskelyne and Devant's.—Past Mysteries Recalled.—The Longest Run in the World," *Observer*, December 10, 1911.

4 **According to Maskelyne**: See Maskelyne, "My Reminiscences."

4 **Henri-Louis Jaquet-Droz**: For a description of the mechanical bullfinch, see "Automata," *Chamber's Journal of Popular Literature, Science, and Art*, February 5, 1876. See also Alfred Chapuis and Edmond Droz, *Automata: A Historical and Technological Study* (London: B. T. Batsford, 1958).

4 **"The delight of seeing"**: Maskelyne, "My Reminiscences."

5 **There are descriptive accounts**: Rau staff, "Famous Automatons and their Rich History," rauantiques.com, June 11, 2021.

5 **There are surviving examples**: Nicholas Reeves, "A Rare Mechanical Figure from Ancient Egypt," *Metropolitan Museum Journal* 50 (2015).

5 **Sumatran statuettes**: For more information on the art of Indonesian puppet theater, see "The History of Indonesian Puppet Theater (Wayang)," education.asianart.org.

5 **Dozens and dozens of such creations**: For a broad history of automata, see Mary Hillier, *Automata and Mechanical Toys: An Illustrated History* (London: Jupiter, 1976). For information on ancient robotics and automata, see Adrienne Mayor, *Gods and Robots: Myths, Machines, and Ancient Dreams of Technology* (Princeton: Princeton University Press, 2018). For information on a praying automaton called the Monk from the sixteenth century, see Elizabeth King and W. David Todd, *Miracles and Machines: A Sixteenth-Century Automaton and Its Legend* (Los Angeles: Getty Publications, 2023). For more information on nineteenth-century automata, see Christian Bailley and Sharon Bailley, *Automata: The Golden Age 1848–1914* (London: Robert Hale, 2003). See also Adelheid Clara Voskuhl, "The Mechanics of Sentiment: Music-Playing Automata and the Culture of Affect in Late Eighteenth-Century Europe," PhD dissertation, Cornell University, 2007.

5 **The work of Pierre and Henri-Louis Jaquet-Droz**: For more on the Jaquet-Droz automata, see the website of the Musée d'Art et d'Histoire in Neuchâtel.

5 **modern-day YouTube videos**: Documentary produced by the Jaquet-Droz watch company about its own history, viewable on YouTube. The section from 4:05 to 7:23 demonstrates the abilities of the three automata.

6 **"'Zoe' bears no resemblance"**: "A Remarkable Automation."

7 **the Turk**: For more on the Turk, see Simon Schaffer, "Enlightened Automata," in William Clark, Jan Golinski, and Simon Schaffer, eds., *The Sciences in Enlightened Europe* (Chicago: University of Chicago Press, 1999). For Maskelyne's thoughts on the Turk, see John Nevil Maskelyne, "Automata," *Leisure Hour*, March 29, 1879.

7 **John Algernon Clark**: Some contemporaneous accounts used the spelling Clarke instead of Clark. Clark appears correct. For a newspaper article crediting Clark as co-inventor of Psycho, see "An Automaton Card-Player: A Machine Which Plays a Good Game of Whist and Generally Wins," *The Times*, February 13, 1875.

7 **"mysterious power of intelligence"**: Ibid.

8 **"The notes are not quite so pure"**: "Fanfare," *Chicago Daily Tribune*, June 16, 1878. This article was reprinted from *The Times*.

8 **"Labial's observance"**: A. J. Phasey, *Musical World*, quoted in Maskelyne, "Automata."

8 **Maskelyne and Cooke billed their act**: "Maskelyne and Cook's Automata."

8 **"Whether Messrs. Maskelyne and Cooke"**: "Messrs. Maskelyne And Cooke."

8 **When they first started playing**: For more on Egyptian Hall in London, see arthur loyd.co.uk.

9 **the Bowery in New York**: For more information on this colorful era of dime-museum entertainment in New Nork, see Andrea Stulman Dennett, *Weird and Wonderful: The Dime Museum in America* (New York: NYU Press, 1997).

9 **"longest run of any entertainment"**: "A Master of Illusion."

9 **A wave of mechanical-people acts**: Zutka, the Mysterious: "Zutka, The Mysterious," *New Zealand Evening Post*, December 3, 1904. Weston the Walking Auto-mation: "Amusements," *Trenton Times*, October 20, 1894. Adam Ironsides and the Steam Man: Cyrenius C. Roe, "Steam Man or Walking Machine," Canadian Patent Number 4175, January 1, 1874; "A Steam Man," *Lebanon Daily News*, September 21, 1880; "The Walking Steam Man," *Reading Times and Dispatch*, August 5, 1878. Phroso the Mechanical Doll: "Doll Shakes Hands with the Audience," *Los Angeles Herald*, March 21, 1906. Ali the Wondrous Electrical Automation: "An Electric Automaton," *Argus* (Australia), August 20, 1887. Moto-Phoso: "Automaton or Actor—Which?," *Popular Electricity: In Plain English* 4, July 1911. Fontinelle the Auto-Man: Orpheum Theater advertisement, *Pensacola Journal*, April 25, 1909. Enigmarelle: "A Clever Mechanical and Electrical Automaton," *Scientific American* 94, no. 2 (January 13, 1906). Perew the Electric Man: W. B. Northrop, "An Electric Man," *Strand Magazine*, November 1900.

9 **"La Motogirl, Célèbre Poupée Electrique"**: Promotional poster, 1905.

10 **In a publicity photo**: *Advertiser* (Adelaide, Australia), November 24, 1903.

11 **Frederic Melville**: For more on Frederic Melville and Doris Chertney, see M. Dinorben Griffith, "The Automaton Girl," *Strand Magazine*, January/June 1905.

11 **"a wonderful mechanical contrivance"**: "Jests at the Continuous: Notes of Plays and Players," *New York Times*, February 27, 1903.

11 **"I will bring the figure"**: Ibid.

11 **"Certainly," he said**: Griffith, "The Automaton Girl."

11 **Touring America**: For thorough coverage of vaudeville acts featuring purported automated men or women, see Steve Carper, "Is It Mechanism or Soul?," *Robots in American Popular Culture* (Jefferson, NC: McFarland, 2019).

12 **"mechanical marvel"**: Christine Gauvreau, "New Haven Encounters 'La Moto Girl,' 1908," Connecticut Digital Newspaper Project, October 28, 2016.

12 **"and all the scientists marveled"**: "Vaudeville Notes," *Washington Post*, February 8, 1903.

12 **She had been inspected**: Griffith, "The Automaton Girl."

12 **She had been submitted**: "At the New Haven: Valadon and His Wonderful Moto Girl in Attraction Next Week," *Journal and Courier* (New Haven, CT), March 14, 1908.

12 **"It seems almost impossible"**: Ibid.

12 **"She is a very dainty"**: "The Motogirl Makes a Hit with a Chase Audience," *Washington Post*, March 24, 1903.

12 **Doris Chertney**: Some contemporaneous accounts of her story used the spelling Chertsey and other variants. Chertney appears correct.

12 **a profile**: Griffith, "The Automaton Girl."

13 **"one of the best shows"**: "Drama and Music," *Boston Daily Globe*, January 22, 1903.

13 **"This 'what is it?' makes"**: *Detroit Free Press*, September 27, 1902.

13 **"'Is it alive?'"**: "Jests at the Continuous: Notes of Plays and Players," *New York Times*, February 27, 1903.

CHAPTER 2: THINKING MACHINES

15 **The period we know as the Enlightenment**: For information on the relationship between machine technology, the Industrial Revolution, and colonialism, see Barbara Hahn, *Technology in the Industrial Revolution* (New York: Cambridge University Press, 2020). For insight into women and the Enlightenment, see Carla Hesse, *The Other Enlightenment: How French Women Became Modern* (Princeton: Princeton University Press, 2001).

16 **In European thought**: For a comprehensive look at the conception of bodies as machines in antiquity, see Maria Gerolemou and George Kazantzidis, *Body and Machine in Classical Antiquity* (Cambridge: Cambridge University Press, 2023). For more information on Pythagoreanism, see Christoph Riedweg, *Pythagoras: His Life, Teaching, and Influence*, translated by Steven Rendall (Ithaca, NY: Cornell University Press, 2005).

16 **The Enlightenment philosopher René Descartes**: For more on Descartes, see Stephen Gaukroger, *Descartes: An Intellectual Biography* (Oxford: Oxford University Press, 1995). For more on Descartes's view of the body as a machine, see Betty Powell, "Descartes' Machines," *Proceedings of the Aristotelian Society* 71 (1970).

16 **Fascinated by hydraulic automata**: For more on the hydraulic statues created by the Francini brothers in the French royal gardens and their impact on Descartes's thinking, see Laura C. Balladur, "Dotted Lines and Fountain Diagrams in Descartes's *Treatise on Man*," *Mosaic: An Interdisciplinary Critical Journal* 51, no. 1 (March 2018).

16 **miniature humanlike figures**: A story about Descartes, almost surely apocryphal, describes him constructing a functioning automaton in the form of a small girl. While Descartes and his creation were on board a ship, the captain found the automaton. Alarmed by its mysterious and possibly ungodly power, the captain threw the mechanical girl overboard. For a consideration of the origins and impact of this story, see Minsoo Kang, "The Mechanical Daughter of René Descartes: The Origin and History of an Intellectual Fable," *Modern Intellectual History* 14, no. 3 (2017).

16 *Treatise on Man*: René Descartes, "The World and Treatise on Man," in *The*

Philosophical Writings of Descartes, vol. 1, translated by John Cottingham, Robert Stoothoff, and Dugald Murdoch (Cambridge: Cambridge University Press, 1985).

16 **The Description of the Human Body**: René Descartes, "Description of the Human Body," in *The Philosophical Writings of Descartes*, vol. 1.

17 **"I suppose the body . . . wheels"**: Descartes, "The World and Treatise on Man," 99, 108.

17 **they would not be just a nuffin'**: Lyrics of "If I Only Had a Brain," written by Harold Arlen and E. Y. "Yip" Harburg for *The Wizard of Oz*.

17 **Marin Mersenne**: For more on Mersenne, see Philippe Hamou, "Marin Mersenne," *The Stanford Encyclopedia of Philosophy*, edited by Edward N. Zalta, Summer 2022 edition.

17 **"Suppose that we were"**: René Descartes, "To Mersenne, 30 July 1640," in *The Philosophical Writings of Descartes*, vol. 3, *The Correspondence*, translated by John Cottingham, Robert Stoothoff, and Dugald Murdoch (Cambridge: Cambridge University Press, 1991), 149.

18 **Thomas Hobbes**: For more on Hobbes, see Aloysius P. Martinich, *Hobbes* (New York: Routledge, 2005); Tom Sorrell, *Hobbes* (London: Routledge & Kegan Paul, 1986).

18 **"Life is but a motion of limbs"**: Thomas Hobbes, *Leviathan: With Selected Variants from the Latin Edition of 1668*, edited by Edwin M. Curley (Indianapolis and Cambridge: Hackett, 1994), 3.

18 **Julien Offray de La Mettrie**: For more on La Mettrie, see Kathleen Wellman, *La Mettrie: Medicine, Philosophy, and Enlightenment* (Durham, NC: Duke University Press, 1992).

18 **L'Homme machine**: The original title of the English translation of La Mettrie's work was *Man a Machine*. A modern translation by Ann Thomson titles it *Machine Man*. See Julien Offray de La Mettrie, *Machine Man and Other Writings*, translated by Ann Thomson (Cambridge: Cambridge University Press, 1996).

18 **"The human body"**: Ibid., 7.

19 **Histoire naturelle de l'âme**: Ibid., 41–73.

19 **"The soul is merely"**: Ibid., 26.

19 **"subtle and marvelous force"**: Ibid., 28.

19 **"I believe that thought"**: Ibid., 35.

19 **Discours sur le bonheur**: This work is also known as *Anti-Senèque ou le Souverain Bien*, translated as *Anti-Seneca, or the Sovereign God*. It is found under this title in La Mettrie, *Machine Man and Other Writings*, 117–43.

20 **"Look 'round the world . . . mind of man"**: David Hume, *Dialogues Concerning Natural Religion*, edited by Dorothy Coleman (Cambridge: Cambridge University Press, 2007), 19.

21 **"infield worked"**: "Piling Up Victories: Tables Turned on Orioles by the Reds—Buffington a Puzzle, and the Infield Worked Like a Machine—Senators Win from the Athletics, and Portland Defeats Worchester," *Boston Daily Globe*, July 9, 1891.

21 **"The human body"**: Lynden Evans, "Domestic Science: Lesson No. 10—Properties of Foods," *Chicago Daily Tribune*, August 12, 1903.

21 **"kept working like a machine"**: "Say Gregg Is Hard Fighter," *Evening Chronicle* (Spokane, WA), June 14, 1907.

21 **"In very truth"**: Leonard K. Hirshberg, "Man a Perfect Machine: The Human Body Will Stand an Amazing Amount of Injury and Abuse," *Detroit Free Press*, December 10, 1910.

22 **"like a machine that has been driven"**: "'If We Studied Old Age, Life Could Be Prolonged': Dr. I. L. Nascher Laments That Too Little Attention Is Paid to What He Has Named 'Geriatrics' by Physicians and Believes That Life Could Be Made Longer and Its Close Happier," *New York Times*, April 21, 1912.

22 **"Man lives and works"**: "Darrow in Debate Jibes at Human Hope of a Hereafter: Declares Man is a Machine Whose Materials can be Bought at Drugstore for 95 Cents—Pres Gray of Bates College Converts Atheist," *Boston Daily Globe*, March 17, 1927.

22 **Mark Twain, bruised by money problems**: For more on Mark Twain, see Ron Powers, *Mark Twain: A Life* (New York: Free Press, 2005).

22 **a pamphlet titled "What Is Man?"**: For an analysis of this work, see Alexander E. Jones, "Mark Twain and the Determinism of 'What Is Man?'," *American Literature* 29, no. 1 (March 1957). To read the text itself, see Mark Twain, *What Is Man? And Other Essays* (New York: Harper & Brothers, 1917).

22 **"What is man, that thou art mindful of him?"**: In the twenty-first-century edition of the King James Bible, this passage is translated: "what is man that Thou art mindful of him, and the son of man that Thou dost visit him? / For Thou hast made him a little lower than the angels, and hast crowned him with glory and honor. / Thou hast made him to have dominion over the works of Thy hands; Thou hast put all things under his feet . . ." *The Holy Bible, Containing the Old and New Testaments: New King James Version* (Nashville: Thomas Nelson, 1994).

22 **"The Old Man and the Young Man"**: Mark Twain, *What Is Man?*, 1.

23 **"O.M. Man the machine"**: Ibid., 5.

23 **"O.M. To me"**: Ibid., 98.

24 **as his "gospel"**: Mark Twain, *Mark Twain in Eruption: Hitherto Unpublished Pages About Men and Events*, edited by Bernard DeVoto (New York: Harper & Brothers, 1940), 239–40.

24 **in a section of *A Connecticut Yankee***: See Mark Twain, *A Connecticut Yankee in King Arthur's Court* (London: Penguin Classics, 1972).

24 **earlier examples of automated dolls**: See "Lady Seated at Her Piano," an automaton created by Gustave Vichy in 1878. The lady has a moving head and hands to mimic playing the instrument, but the music is created by a music box hidden in the body of the piano.

26 **one in every six households**: "The Piano Industry," *Wall Street Journal*, January 14, 1905.

26 **"The extent to which pianos"**: Ibid.

26 **majority . . . were female**: For more on the relationship between femininity and piano playing in the early twentieth century, see Francesca Carnevali and Lucy Newton, "Pianos for the People: From Producer to Consumer in Britain, 1851–1914," *Enterprise and Society* 14, no. 1 (March 2013).

26 **Otto Ortmann**: For more on Otto Ortmann, see Dale Keiger, "Piano Playing as Science," *Johns Hopkins Magazine*, April 2000.

27 ***The Physiological Mechanics of Piano Technique***: Otto Ortmann, *The Physiological Mechanics of Piano Technique: An Experimental Study of the Nature of Muscular Action as Used in Piano Playing and of the Effects Thereof Upon the Piano Key and the Piano Tone* (London: Kegan Paul, Trench, Trubner, 1929).

27 **"Since the final end"**: Ibid., 3.

27 **The principle of the parallelogram"**: Ibid., 10.

27 **"Otto Ortmann and Rudolf Ganz"**: Author's interview with Sophia Rosoff.

27 **Sophia Rosoff**: For more on Sophia Rosoff and Abby Whiteside, see the Whiteside Foundation website.

28 **"I learned from . . . do the job"**: Author's interview with Sophia Rosoff.

CHAPTER 3: MORE COULD NOT BE ASKED OF MORTAL INGENUITY

32 **Boxes that made music**: See Q. David Bowers, *Encyclopedia of Automatic Musical Instruments* (New York: Vestal Press, 1972). For information on cylinder music boxes, see 15–95. For information on disc music boxes, see 97–252.

32 **An "orchestra box"**: Ibid., 343–736.

32 **the cylinders would play Mendelssohn's**: See Arthur W. J. Ord-Hume, *Clockwork Music: An Illustrated History of Mechanical Musical Instruments from the Musical Box to the Pianola, from Automaton Lady Virginal Players to Orchestrion* (New York: Crown, 1973).

33 **"Many people are forced"**: Ord-Hume, *Clockwork Music*, 238–39.

33 **Exceedingly elaborate music-playing mechanisms**: For background information, see Bowers, *Encyclopedia of Automatic Musical Instruments*; Cynthia Hoover, *The History of Musical Machines* (New York: Drake, 1975).

34 **An exemplary specimen**: Details on the Popper Rex Orchestrion and the Guinness Collection of Mechanical Music and Automata at the Morris Museum come from the author's visit to the museum. For more on the Guinness Collection, see morris museum.org.

34 **the Popper company**: For more on the Popper company, see Q. David Bowers, "Popper & Company—A Musical Dynasty," *The Music Box: The Journal of the Musical Box Society of Great Britain* 5, no. 7 (December 1972).

34 **the Regina company**: For more on the Regina company's music boxes and sheets, see Larry Karp, "Regina and Ragtime," *The Music Box: The Journal of the Musical Box Society of Great Britain* 59, no. 2 (2013).

35 **"Hallelujah"**: For more on Vincent Youmans, see Gerald Martin Bordman, *Days to Be Happy, Years to Be Sad: The Life and Music of Vincent Youmans* (New York: Oxford University Press, 1982). See also his definitive collection: Vincent Youmans, *Music of Vincent Youmans* (New York: Columbia, 1951).

36 **conventional pianos were firmly established**: For more on the history of pianos, see James Parakilas, *Piano Roles: Three Hundred Years of Life with the Piano* (New Haven: Yale University Press, 1999); Robert Palmieri and Margaret W. Palmieri, eds., *Piano: An Encyclopedia* (New York: Routledge, 2003).

37 **player pianos**: For more, see Brian Dolan, *Inventing Entertainment: The Player Piano and the Origins of an American Musical Industry* (Lanham, MD: Rowman & Littlefield, 2009); Francis D. Klingender, *Art and the Industrial Revolution* (Gloucester, UK: Royle, 1947).

37 **the Aeolian company's Pianola**: For more, see Cynthia Adams Hoover, "Aeolian Corporation," Oxford Music Online, July 1, 2014.

37 **By 1915**: Bowers, *Encyclopedia of Automatic Musical Instruments*, 256.

37 **By 1920**: Ibid.

37 **"The Pianola is found"**: Ibid., 255.

37 **"It is hardly possible"**: Ibid., 258.

38 **Player-piano music**: For more on the player piano, see Allison Rebecca Wente, *The Player Piano and Musical Labor: The Ghost in the Machine* (Abingdon, UK: Routledge, 2022).

38 **An early article**: Wm. Strauss, "How Automatic Piano Player is Made to Produce Music," *Chicago Daily Tribune*, July 15, 1906.

38 **M. E. Brown of the U.S. Music Company**: For more on M. E. Brown, see Artis Wodehouse, "Popular US Women Piano Roll Artists 1910–1930, a Scrapbook," academia .edu.

38 **George Gershwin's hit song "Swanee"**: For more on this song, see Paul Zollo, "Legends of Songwriting: Irving Caesar, the Guy who wrote 'Swanee' with Gershwin," *American Songwriter*, 2020.

39 **Gershwin himself**: For more, see Artis Wodehouse, "Tracing Gershwin's Piano Rolls," in Wayne Schneider, ed., *The Gershwin Style: New Looks at the Music of George Gershwin* (Oxford: Oxford University Press, 1999).

39 **Igor Stravinsky made**: For more, see Mark McFarland, "Stravinsky and the Pianola: A Relationship Reconsidered," *Revue de Musicologie*, no. 1 (2011).

39 **"struck on the head with a hickory club"**: "Well Trained Ear Suffers: Musician Declares Automatic Piano Is Nuisance," *Los Angeles Times*, September 26, 1909.

40 **"Altogether the artificial"**: "Artist and Mechanic Agree: Automatic Piano Player Accepted by All as Factor in Musical Progress," *Christian Science Monitor*, August 19, 1911.

40 **"people of culture"**: "How the Untrained Fingers of Business Men and Blacksmiths Render Stately Music of the Masters," *San Francisco Chronicle*, August 14, 1904.

40 **Ragtime was a form**: For more on the history of ragtime, see David A. Jasen and Trebor Jay Tichenor, *Rags and Ragtime: A Musical History* (New York: Dover, 1989); Rudi Blesh, *They All Played Ragtime: The True Story of an American Music* (London: Sidgwick & Jackson, 1958); and John Edward Hasse, ed., *Ragtime: Its History, Composers, and Music* (New York: Schirmer, 1985).

41 **By the mid-1920s, piano rolls were selling**: Michael Montgomery, Trebor Jay Tichenor, and John Edward Hasse, "Ragtime on Piano Rolls," in Hasse, ed., *Ragtime*.

41 **Eubie Blake**: See Al Rose, *Eubie Blake* (New York: Schirmer, 1979); Richard Carlin, *Eubie Blake: Rags, Rhythm, and Race* (New York: Oxford University Press, 2020).

41 **James P. Johnson**: See Tom Davin, "Conversations with James P. Johnson," in Hasse, ed., *Ragtime*.

42 **Conlon Nancarrow**: See Jürgen Hocker, *Encounters with Conlon Nancarrow* (Lanham, MD: Lexington, 2012); Kyle Gan, *The Music of Conlon Nancarrow* (Cambridge: Cambridge University Press, 1995). Some material in this section first appeared in David Hajdu, "Look, No Hands! The Impossibly Original Music of Conlon Nancarrow," *Oxford American*, Summer 1998.

42 **"dreaming of getting rid of the performers"**: Charles Amirkhanian, "Interview with Composer Conlon Nancarrow," *Soundings*, Spring/Summer 1977.

42 **Henry Cowell**: Nancarrow's thinking on music was heavily influenced by Cowell, who advocated "writing music specially for player-piano." Henry Cowell, *New Musical Resources* (New York: Alfred A. Knopf, 1930), 64–65.

42 **"I was always constrained"**: John Rockwell, "Conlon Nancarrow: Poet of the Player Piano," *New York Times*, June 28, 1981.

43 **"All those keys"**: Hajdu, "Look, No Hands!"

43 **"Most of my things"**: James R. Greeson et al., "Conlon Nancarrow: An Arkansas Original," *Arkansas Historical Quarterly* 54, no. 4 (1995).

43 **Nancarrow was discovered**: See Greeson, "Conlon Nancarrow"; Roger Reynolds,

"Conlon Nancarrow: Interviews in Mexico City and San Francisco," *American Music* 2, no. 2 (1984); David Bruce, "The Manic Mechanic," *Musical Times* 138, no. 1850 (1997).

43 **"His music is so utterly original"**: Letter from György Ligeti to Charles Amirkhanian, quoted in Gan, *The Music of Conlon Nancarrow*, 2.

44 **"Music is for listening"**: Rockwell, "Conlon Nancarrow: Poet of the Player Piano."

44 **The flute has some 120 parts**: See Sian Hughes, "The Different Parts of a Flute: Anatomy and Structure," *Hello Music Theory*, August 4, 2022.

44 **"Guthrie began to think"**: Michael J. Kramer, "This Machine Kills Fascists: Technology and Folk Music in the USA." Work in progress.

45 **"The piano is a machine"**: "Odd House That Follows the Sun: Josef Hofmann, Pianist, Describes the Dwelling He Planned Several Years Ago—Musicians, He Says, Often Are Good Mechanics," *New York Times*, May 22, 1927.

45 **"The piano is a percussion . . . what I'm saying?"**: Author's interview with Donald Shirley.

46 **"I play a little differently"**: Author's interview with Herbie Hancock.

CHAPTER 4: EVEN THE KITCHEN SINK

49 **Machine Age Exposition**: For more on this, see Barnaby Haran, "Constructivism in the USA: Machine Art and Architecture at the Little Review Exhibitions," *Watching the Red Dawn: The American Avant-garde and the Soviet Union* (Manchester: Manchester University Press, 2016); Herbert Lippmann, "The Machine-Age Exposition," *The Arts*, June 1927.

50 **a Model M**: For more on the history of Steinway pianos, see Theodore E. Steinway, *People and Pianos: A Pictorial History of Steinway & Sons* (Pompton Plains, NJ: Amadeus Press, 2005); Richard K. Lieberman, *Steinway and Sons* (New Haven: Yale University Press, 1995). For information on the building of a Steinway piano, see this documentary: steinway.com/misc/note-by-note.

50 **The Little Review**: See Margaret Anderson, ed., *The Little Review Anthology* (New York: Hermitage House, 1953).

50 **"That the machine"**: E. B. White, "Machine Age," *The New Yorker*, May 21, 1927.

50 **Jane Heap**: For more, see Linda Lappin, "Jane Heap and Her Circle," *Prairie Schooner* 78, no. 4 (2004). For information on both Jane Heap and Margaret Anderson, see Holly A. Baggett, *Making No Compromise: Margaret Anderson, Jane Heap, and the "Little Review"* (DeKalb, IL: Northern Illinois University Press, 2023).

51 **Margaret Anderson**: See Margaret Anderson's autobiography in three volumes: Margaret C. Anderson, *My Thirty Years' War: An Autobiography* (New York: Covici, Friede, 1930); Margaret C. Anderson, *The Fiery Fountains: The Autobiography: Continuation and Crisis to 1950* (New York: Hermitage House, 1951); and Margaret C. Anderson, *The Strange Necessity* (New York: Horizon Press, 1970). See also Jackson R. Bryer, *A Trial-track for Racers: Margaret Anderson and the Little Review* (Madison: University of Wisconsin, 1965).

51 **"From the insane"**: Jane Heap, "Mary Garden," *Little Review*, March 1917.

51 **"my blessed antagonistic complement"**: Jane Heap, letter to Florence Reynolds, March 17, 1918. Holly A. Baggett, ed., *Dear Tiny Heart: The Letters of Jane Heap and Florence Reynolds* (New York: NYU Press, 2000), 55.

51 **"To express the emotions"**: Heap, "Mary Garden."

51 **Anderson and Heap went to trial**: For more on this trial, see Edward de Grazia,

Girls Lean Back Everywhere: The Law of Obscenity and the Assault on Genius (New York: Random House, 1992), ch. 1.

52 **"There is a great new race ... present"**: Jane Heap, "Machine-Age Exposition," *Machine-Age Exposition Catalogue*, May 1928.

52 **George Gurdjieff**: For more on George Gurdjieff, see Margaret C. Anderson, *The Unknowable Gurdjieff* (New York: S. Weiser, 1962). Anderson dedicated this book to Jane Heap.

52 **"the work"**: For more on "the work," see P. D. Ouspensky, *In Search of the Miraculous* (New York: Harcourt, Brace, Jovanovich, 1977).

52 **"Every one of you"**: George Gurdjieff, *Views from the Real World: Early Talks in Moscow, Essentuki, Tiflis, Berlin, London, Paris, New York, and Chicago* (New York: E. P. Dutton, 1973), 49.

53 **the *Futuristi***: See Katia Pizzi, *Italian Futurism and the Machine* (Manchester: Manchester University Press, 2019); Walter L. Adamson, "How Avant-Gardes End—and Begin: Italian Futurism in Historical Perspective," *New Literary History* 41, no. 4 (Autumn 2010); and Mark A. Radice, "'Futurismo': Its Origins, Context, Repertory, and Influence," *Musical Quarterly* 73, no. 1 (1989).

53 **series of self-defined manifestos**: The earliest work, *The Founding and Manifesto of Futurism*, was published in 1909. The next two, *Manifesto of Futurist Painters* and *Manifesto of Futurist Musicians*, begin to set forward the principles of Futurism with regard to the arts.

53 **anticipated the rise of Italian Fascism**: For more on the relationship between Futurism and Fascism, see Anne Bowler, "Politics as Art: Italian Futurism and Fascism," *Theory and Society* 20, no. 6 (1991); Simonetta Falasca-Zamponi, "The Artist to Power? Futurism, Fascism and the Avant-Garde," *Theory, Culture & Society* 13, no. 2 (May 1996). For information on the tenets of futurism and an evisceration of pasta, see Filippo Tommaso Marinetti, *The Futurist Cookbook*, edited by Lesley Chamberlain, translated by Suzanne Brill (New York: Penguin Modern Classics, 2014).

53 **"Is not the machine today"**: Enrico Prampolini, "The Aesthetic of the Machine and Mechanical Introspection in Art," *Machine-Age Exposition Catalogue*, May 1928.

54 **"Machines themselves"**: Herbert Lippmann, "The Machine-Age Exposition," *The Arts*, June 1927.

55 **"It may be that the machine"**: Genevieve Taggard, "The Ruskinian Boys See Red," *New Masses*, July 1927.

56 **"If women would once try flying"**: "Urges Women Take Up Flying," *Boston Daily Globe*, June 26, 1927.

57 **"As a matter of fact"**: "Machine Art," press release, Museum of Modern Art, March 5, 1934.

57 **"Machine Art"**: For more, see Jennifer Jane Marshall, *Machine Art, 1934* (Chicago: University of Chicago Press, 2012).

58 **"There are no"**: Philip Johnson, "History of Machine Art," *Machine Art: March 6 to April 30, 1934* (New York: Museum of Modern Art, 1934).

59 **"In spirit machine art"**: Ibid.

59 **"By beauty of shapes"**: Ibid.

59 **In their coverage**: Edward Alden Jewell, "Art Scans Its Niche in Industrial Plan," *New York Times*, February 26, 1934; Walter Rendell Storey, "Machine Art Enters the Museum Stage," *New York Times*, March 4, 1934; Edward Alden Jewell, "Machine Art Seen in Unique Exhibit," *New York Times*, March 6, 1934; Royal Cortissoz, "Machine Art and the Art of Some Artists," *New York Herald Tribune*, March 1,

1934; Joseph W. Alsop, "Pots and Sinks Going on View as Art at Machinery Exhibit," *New York Herald Tribune*, March 5, 1934; Arthur Millier, "Is Beauty the Same Thing in Auto's Lines and in Sunset?," *Los Angeles Times*, March 25, 1934; and E. C. Sherburne, "Even the Kitchen Sink," *Christian Science Monitor*, March 10, 1934.

60 **Plato considered geometry**: For more, see A. Philip McMahon, "Would Plato Find Artistic Beauty in Machines?," *Parnassus* 7, no. 2 (1935).

60 **"Already the machine"**: "The Art of the Machine," *Chicago Daily Tribune*, May 27, 1934.

60 **"It's disturbing"**: Robert M. Coates, "Machine Art," *The New Yorker*, March 17, 1934.

60 **"The work of art presupposes"**: Cortissoz, "Machine Art and the Art of Some Artists."

61 **"to pursue outside interests"**: Alan R. Blackburn, Jr., papers in the Museum of Modern Art Archives, 1992.

61 **Johnson wrote unbridled**: See Kazys Varnelis, "'We Cannot Not Know History': Philip Johnson's Politics and Cynical Survival," *Journal of Architectural Education* 49, no. 2 (1995); Michael Sorkin, "Where Was Philip?," *Spy*, October 1988.

61 **Johnson would eventually come out**: Franz Schulze, *Philip Johnson: Life and Work* (New York: Knopf, 1994).

61 **"You simply could not fail"**: Ibid., 89.

61 **an admiring review of *Mein Kampf***: Philip Johnson, "Mein Kampf and the Businessman," *Examiner*, Summer 1939.

CHAPTER 5: SPIRITUALS OF THE CITY

63 **the microphone**: See John Woram, "50 Years of Studio Mikes," *Billboard*, December 18, 1976. .

63 **As with all electronic machinery**: For more on early electronics, see G. W. A. Dummer, *Electric Inventions and Discoveries: Electronics from Its Earliest Beginnings to the Present Day,* 3rd ed. (New York: Pergamon Press, 1983).

64 **"One thing that was tremendously important"**: Frank Sinatra, "Me and My Music," *Life*, March 18, 1965.

65 **Lee de Forest**: Now largely forgotten, Lee de Forest was once revered as an epoch-making innovator in electronic technology. Generally credited with the invention of the vacuum tube, he was called the "Father of Radio" in his time. For more, see James A. Hijiya, *Lee de Forest and the Fatherhood of Radio* (Bethlehem, PA: Lehigh University Press, 1992); Michael H. Adams, *Lee de Forest: King of Radio, Television, and Film* (New York: Copernicus, 2012); Georgette Carneal, *A Conqueror of Space: An Authorized Biography of the Life and Work of Lee de Forest* (New York: H. Liveright, 1930).

65 **"She was an instant success"**: Lee de Forest, *Father of Radio: The Autobiography of Lee de Forest* (Chicago: Wilcox & Follett, 1950), 351.

65 **"A flood of fan mail . . ."**: Ibid.

65 **"the Radio Girl"**: Irving Settel, *A Pictorial History of Radio* (New York: Grosset & Dunlap, 1967), 58.

66 **"the First Lady of Radio"**: de Forest, *Father of Radio*, 351.

66 **"He's So Unusual"**: Composed by Tin Pan Alley songwriters Avrum Sherman, Alvin Lewis, and Abner Silver, this song was recorded in 1929 by Helen Kane, the voice of Betty Boop in sexually daring cartoons of the late Jazz Age. With its lyrics about a handsome man uninterested in women, the song was embraced by the underground

gay scene of the day. For an overview of early queer music, see George Chauncey, *Gay New York* (New York: Basic Books, 1994) and the Queer Music History website. In 1983, Cindy Lauper played off the transgressive theme of the song on her debut album, *She's So Unusual*, on which the original song figures in multiple tracks.

66 **"'Original Radio Girl' Won Fame"**: Israel Klein, "'Original Radio Girl' Won Fame With Crooning Voice," *Miami Daily News–Record*, May 13, 1930.

67 **"I desire to speak earnestly"**: "Crooning Not Art, Just Whining, Slush, Asserts Cardinal O'Connell," *New York Herald Tribune*, January 11, 1932.

67 **Executives of radio companies**: Daniel Goldmark, "'Making Songs Pay': Tin Pan Alley's Formula for Success," *Musical Quarterly* 98, no. 1–2 (2015).

67 **"The male star prototype softened"**: Ian Whitcomb, "Introduction: The Coming of the Crooners," in Michael Pitts, Frank Hoffmann, Dick Carty, and Jim Bedoian, *The Rise of the Crooners: Gene Austin, Russ Columbo, Bing Crosby, Nick Lucas, Johnny Marvin, and Rudy Vallee* (Lanham, MD: Scarecrow Press, 2001), 1–51.

67 **"twentieth-century means for making music"**: John Cage, "The Future of Music: Credo," in *Silence: Lectures and Writings* (Middletown, CT: Wesleyan University Press, 1961), 6. According to Cage's introduction, the text was initially presented as a "talk at a meeting of a Seattle arts society organized by Bonnie Bird in 1937," and first printed in liner notes to a live recording of Cage's music in 1958.

67 **Seagulls squawked**: The Shangri-Las, "Remember (Walking in the Sand)," written and produced by George "Shadow" Morton, 1965.

68 **Chickens clucking**: This moment occurs at the transition from "Good Morning" to the reprise of the title song. The Beatles, "Good Morning Good Morning," written by John Lennon and Paul McCartney, produced by George Martin, on *Sgt. Pepper's Lonely Hearts Club Band*, 1967.

68 **"Stephen [Stills] and Graham [Nash] and I"**: Author's interview with David Crosby.

69 **Avery Fisher**: For more, see Dave Marsh, "A Conversation with Avery Fisher," *Rolling Stone*, September 23, 1976. See also Avery Fisher, interview by Barbaralee Diamonstein-Spielvogel, 1979, Barbaralee Diamonstein-Spielvogel Collection, David M. Rubenstein Rare Book and Manuscript Library, Duke University.

69 **"My God"**: Author's recollection of the event, Consumer Electronics Show, Chicago, June 1981.

69 **more than two million Black Americans**: This number grew steadily and reached six million by midcentury. See "The Great Migration (1910–1970)," archives.gov, 2021. For more on the Great Migration, see also Isabel Wilkerson, *The Warmth of Other Suns* (New York: Random House, 2010); Nicholas Lemann, *The Promised Land: The Great Black Migration and How It Changed America* (New York: Knopf, 1991).

69 **"I followed the money . . . this place"**: Author's interview with Honeyboy Edwards for "Who's Got the Blues?," *Mother Jones*, September/October 2003. Some of this material did not appear in the article.

70 **electrified models with solid wood bodies**: For more on the electric guitar, see André Millard, ed., *The Electric Guitar: A History of an American Icon* (Baltimore: Johns Hopkins University Press, 2004).

71 **"When I went into the clubs"**: James Rooney, *Bossmen: Bill Monroe & Muddy Waters* (New York: Dial, 1971), 112.

71 **"South Side clubs"**: Francis Davis, *The History of the Blues: The Roots, the Music, the People from Charley Patton to Robert Cray* (New York: Hyperion, 1995), 181.

71 **"It seems as though"**: Richard Wright, *12 Million Black Voices: A Folk History of the Negro in the United States* (New York: Viking, 1941), 109.

71   **"The blues could be called"**: Richard Wright, "Note Sur les Blues," *La Revue du Jazz*, April 1949.

72   **Alan Lomax and John Wesley Work III**: For examples of Lomax's work, see Alan Lomax, *The Land Where Blues Began* (New York: Pantheon, 1993); Alan Lomax, *Folk Song Style and Culture* (Washington DC: American Association for the Advancement of Science, 1968). For more on Lomax himself, see John Szwed, *Alan Lomax: The Man Who Recorded the World* (New York: Viking Penguin, 2010); David Hajdu, "Authenticity Blues," *New Republic*, June 16, 2002.

72   **"We excited the people"**: Author's interview with Honeyboy Edwards.

73   **"A lot of things changed when we got up North"**: Author's interview with James Cotton for "Who's Got the Blues?," *Mother Jones*, September/October 2003. This material did not appear in the article.

74   **"Rocket 88"**: This record was recorded and released in 1951. The song became a hit, climbing quickly to the top of the R & B charts. Jackie Brenston was the singer, performing with Ike Turner and his Kings of Rhythm, although the record identified the group as Jackie Brenston and his Delta Cats.

74   **"Ambition is a dream"**: This quote is widely attributed to Elvis Presley, though the original source proves elusive.

74   **"The sound of the airplane"**: Roger McGuinn, liner notes to the Byrds, *Mr. Tambourine Man*, Columbia Records, 1965.

75   **"We got trippy, man"**: Author's interview with David Crosby.

75   **River Rouge**: For more on the history of the River Rouge complex, see the website of The Henry Ford historical organization, thehenryford.org.

75   **"I remember being led up"**: Mike Flaherty, "The Survivors: Iggy Pop," *GQ*, November 9, 2011.

75   **"Whewwwww!"**: Iggy Pop, in *Motor City's Burning: Detroit from Motown to the Stooges, Part 3*, BBC, 2008.

76   **"Detroit has a beat"**: Brian McCollum, "Russ Gibb—Detroit rock visionary, schoolteacher and 'Paul is dead' prankster—dies at 87," *Detroit Free Press*, April 30, 2019.

76   **"Those slow-moving car frames"**: Berry Gordy, *To Be Loved: The Music, the Magic, the Memories of Motown: An Autobiography* (New York: Warner Books, 1994), 69.

76   **"My own dream"**: Ibid., 140.

76   **"my first musical gift from God"**: Chris Richards, "Can Iggy Pop Hear the Future?," *Durango Herald*, September 11, 2019.

77   **"Before we were making records"**: "Iggy Pop: 'What Happens When People Disappear'," *Morning Edition*, NPR, May 2, 2013.

77   **"driving monotony and terrifying intensity"**: Billy Hough, "John Cale: The Midwife of Punk Rock," pleasekillme.com, March 7, 2019.

77   **"Not much in that music speaks"**: Author's interview with John Cale.

CHAPTER 6: THIS IS MUSIC?

79   **the Illiac**: For more on this moment in computer history, see Herman Heine Goldstine, *The Computer from Pascal to von Neumann* (Princeton: Princeton University Press, 1972).

80   **Dr. Lejaren A. Hiller Jr.**: See James Bohn, "Lejaren Hiller: Early Experiments in Computer and Electronic Music," in Lillian Hoddeson, ed., *No Boundaries: University of Illinois Vignettes* (Champaign: University of Illinois Press, 2004); Allan Kozinn, "Lejaren Hiller, 69, First Composer To Write Music With a Computer,"

New York Times, February 1, 1994. The University at Buffalo compiled an archive of articles written about Hiller, as well as Hiller's papers and correspondence.

80 **"I had an idea one day"**: Lejaren A. Hiller Jr. interviewed by Vincent Plush, November 12, 1983, for "Major Figures in American Music," Oral History of American Music, Yale University Music Library. This quote was used by John Bewley in his exhibition catalog for "Lejaren A. Hiller: Computer Music Pioneer," at SUNY Music Library, University at Buffalo, May 24–September 7, 2004. The quote also appears in Tiffany Funk, "A Music Suite Composed by an Electronic Brain," *Leonardo Music Journal* 28 (2018).

80 **Lejaren à Hiller**: See "Using the Camera to Illustrate Fiction: Models Pose for Photographs Showing Scenes in the Story—How Two Artists Originated the Plan," *New York Times*, January 6, 1918. To view the elder Hiller's reconstructions of historical surgeries, see Lejaren à Hiller, *Surgery Through the Ages: A Pictorial Chronical* (New York: Hastings House, 1944).

80 **"wild parties with nude women"**: Hiller interviewed by Plush.

81 **"cutting designs . . . effects"**: David Ewen, *American Composers: A Biographical Dictionary* (New York: G. P. Putnam's Sons, 1982), 328.

81 **"a bootleg job"**: Hiller interviewed by Plush.

81 **"Music is a sensible form"**: Lejaren A. Hiller Jr. and Leonard M. Isaacson, *Experimental Music Composition with an Electronic Computer* (New York: McGraw-Hill, 1959), 2.

82 **"It is a feature of digital computers"**: Ibid.

82 **"It should be noted"**: Ibid.

83 **"a complete departure"**: Ibid., 5.

83 **"had a very electric atmosphere"**: Hiller interviewed by Plush.

83 **"This Is Music?"**: William D. Hansen, "This is Music?," *Chicago Daily Tribune*, April 12, 1959.

84 **"To stretch the point"**: Brock Bower, "These Thinking Machines Can't Really Think: Electronic Brains Do Some Things Far Better Than the Old Fashioned Human Kind, But, We Are Assured, Homo Sapiens Need Not Yet Head for the Hills," *New York Times*, April 9, 1961.

84 **"It was really a very strange summer . . . really"**: Hiller interviewed by Plush.

84 **"conversing with the computer"**: Hiller and Isaacson, *Experimental Music Composition*, 65.

84 **"Being 'creative'"**: Ibid., 68.

85 **Brainiac**: The DC character Brainiac, created by writer Otto Binder and artist Al Plastino, appeared for the first time in the July 1958 issue of *Action Comics*, although his identity as an android was not yet established. Three months earlier in a special issue of *Superman's Pal, Jimmy Olsen*, Binder introduced a separate character named Brainiac, a robotic "electronic brain" from the sixtieth century.

85 **Ada Byron, Countess of Lovelace**: See Christopher Hollings, Ursula Martin, and Adrian Rice, *Ada Lovelace: The Making of a Computer Scientist* (Oxford: Bodleian Library, 2018); Doris Langley Moore, *Ada: Countess of Lovelace* (London: John Murray, 1977); Joan Baum, *The Calculating Passion of Ada Byron* (Hamden, CT: Archon Books, 1986); Ada Lovelace, *Ada, the Enchantress of Numbers: A Selection from the Letters of Lord Byron's Daughter and Her Description of the First Computer*, edited by Betty A. Toole (Mill Valley, CA: Strawberry Press, 1992).

86 **"[The Engine] might act upon other things"**: Ada Byron, Countess of Lovelace, translated a paper on Babbage's Engine by the mathematician (and seventh prime

minister of Italy) Luigi Federico Menabrea. Lovelace published this translation along with her own significantly longer notes. See L. F. Menabrea, "Sketch of the Analytical Engine invented by Charles Babbage, Esq.," in Richard Taylor, ed., *Scientific Memoirs, Selected from the Transactions of Foreign Academies of Science and Learned Societies, and from Foreign Journals*, vol. 3 (London: Richard and John E. Taylor, 1843), 694.

86   **the programmer Betty Snyder**: See Claudia Levy, "Betty Holberton Dies," *Washington Post*, December 11, 2001; W. Barkley Fritz, "The Women of ENIAC," *IEEE Annals of the History of Computing* 18, no. 3 (1996). For more on the role of women in early computing see Jennifer S. Light, "When Computers Were Women," *Technology and Culture*, July 1999.

87   **"singing songs and stuff"**: Jean J. Bartik, in Jean J. Bartik and Frances E. Snyder Holberton interviewed by Henry S. Tropp, April 27, 1973, transcript, Archives Center, National Museum of American History, 169. There is some confusion about the date of this event. Snyder is reported to have said in an earlier conversation that the BINAC played music in 1948; see Kathryn Kleiman, quoted in "Computer Music in 1949?," Irrlicht Project, November 11, 2015. In the Tropp interview, Snyder and Bartik agree that this event took place in the spring of 1949.

87   **"banal tune-maker"**: R. C. Pinkerton, "Information Theory and Melody," *Scientific American* 194 (February 1956).

87   **"The liberal arts"**: David Sarnoff, liner notes to boxed set of 7-inch, 45-RPM records, *The Sounds and Music of the RCA Electronic Music Synthesizer*, RCA Victor, 1955.

88   **There was a cartoon**: This drawing was the work of French illustrator J. J. Grandville, originally published in 1844. See J. J. Grandville, *Un Autre Monde: Transformations, Visions, Incarnations, Ascensions, Locomotions, Explorations, Pérégrinations, Excursions, Stations, Cosmogonies, Fantasmagories, Rêveries, Folatreries, Facéties, Lubies, Metamorphoses, Zoomorphoses, Lithomorphoses, Métempsycoses et Autres Choses* (Paris: H. Fournier, 1844), 17.

89   **Datatron 205**: For more on the Datatron 205 and other Burroughs computers, see G. T. Gray and R. Q. Smith, "Before the B5000: Burroughs computers, 1951–1963," *IEEE Annals of the History of Computing* 25, no. 2 (April/June 2003). For a survey of appearances of the Datatron 205 on screen, see starringthecomputer.com.

89   **Do not skip**: Martin L. Klein, "Syncopation by Automation," *Radio Electronics* 28, no. 6 (June 1957).

89   **"set out to prove"**: Ibid.

90   **"Push Button Bertha"**: Unfortunately, no recording of Jack Owen singing the song is known to have survived.

90   **"The report of a new electronic brain"**: "Musical Automation," *Hartford Courant*, July 6, 1956.

90   **"If the Datatron"**: "1,000 Tunes an Hour," *Austin Statesman*, July 21, 1956.

90   **"If you should happen to be walking"**: "'Miracle Brain' Providing Headache to Songwriters," *Michigan Chronicle*, September 8, 1956.

CHAPTER 7: BRAIN AUTOMATION

93   **"along with their various dialects"**: *Forbidden Planet*, directed by Fred M. Wilcox, MGM, 1956. In this midcentury sci-fi film, the crew of an Earth spaceship arrives on the planet Altair IV, where it is welcomed by a large robot who introduces himself as Robby. With the politeness and dry wit of a Hollywood butler, Robby leads the earthlings to the residence of Dr. Morbius.

94 *Cybernetics*: Norbert Wiener, *Cybernetics: Or, Control and Communication in the Animal and the Machine* (Cambridge, MA: MIT Press, 1948).

94 *Automata Studies*: C. E. Shannon and J. McCarthy, eds., *Automata Studies* (Princeton: Princeton University Press, 1956).

94 academic symposium on "Cerebral Mechanisms": This event, known as the Hixon symposium, took place at the California Institute of Technology in September 1948. A book was published with six of the papers presented at the symposium along with transcriptions of discussions by the participants. See Lloyd A. Jeffress, ed., *Cerebral Mechanisms in Behavior: The Hixon Symposium* (New York: Wiley, 1951).

94 "interacting finite automata," "brain automation," "No one had talked about": Nils J. Nilsson, "John McCarthy: 1927–2011," National Academy of Sciences, 2012.

94 "I was hoping": John McCarthy interviewed by Nils J. Nilsson, September 12, 2007, Oral History Collection, Computer History Museum, Mountain View, CA.

95 "I didn't have the phrase": Ibid.

95 "My idea": John McCarthy interviewed by Wendy Conquest, Dan Rockmore, and Bob Drake, July 26, 2005, Science Lives Series, Simons Foundation, New York.

95 "a 2 month, 10 man study": J. McCarthy, M. L. Minsky, N. Rochester, and C. E. Shannon, "A Proposal for the Dartmouth Summer Research Project on Artificial Intelligence," August 31, 1955, 2.

95 "The study is to proceed": Ibid.

96 "Culver accompanied McCulloch": Trenchard More, *Dartmouth Conference on Artificial Intelligence: Report on the 5th and 6th Weeks*, Dartmouth Summer Research Project on Artificial Intelligence, July 31, 1956, 4.

97 "Certain things . . . meetings": John McCarthy interviewed by Conquest, Rockmore, and Drake.

99 "Descartes, in *De Homine*": Shannon and McCarthy, eds., *Automata Studies*, v.

100 "both species belonging": A. Newell and H. Simon, *Human Problem Solving* (Englewood Cliffs, NJ: Prentice-Hall, 1972), 870.

100 "two kinds of automata": John von Neumann, *The Computer and the Brain* (New Haven: Yale University Press, 1958), 39.

100 "The most immediate observation": Ibid., 40.

101 "the problem of what understanding is": Steven L. Fair, "Man's Bid to Outsmart Himself: The Day May Soon Come When Computerized Artificial Intelligence Gains the Upper Hand," *Boston Globe*, September 24, 1978.

CHAPTER 8: EVERYBODY SHOULD BE A MACHINE

103 "Money corrupts! Art erupts!": *What a Way to Go!*, directed by J. Lee Thompson, 20th Century Fox, 1964.

104 "The sound": Ibid.

104 Betty Comden and Adolph Green: Betty Comden and Adolph Green began their career together in a topical sketch comedy act called the Revuers, which sometimes featured Leonard Bernstein as pianist.

106 "Warhol's currency paintings": Blake Gopnik, *Warhol* (New York: Ecco, 2020), 251.

107 "Everybody should be a machine": Gene R. Swenson, "What Is Pop Art? Answers from 8 Painters, Part I," *Artnews* 62, no. 7 (November 1963).

107 the scholar Jennifer Sichel: See "'Do you think Pop Art's Queer?' Gene Swenson and Andy Warhol," and "'What is Pop Art?' A Revised Transcript of Gene Swenson's 1963 interview with Andy Warhol," *Oxford Art Journal* 41, no. 1 (March 2018).

108 **"Mechanical means are today"**: Swenson, "What Is Pop Art?"

108 **"This crude transfer"**: Gopnik, *Warhol*, 268.

108 **"blank, blunt, bleak, stark"**: Ibid., 218. The full quotation reads, "The idea of doing blank, blunt, bleak, stark images like his was contrary to the whole prevailing mood of the arts—but it gave me a chill."

108 **"The rubber-stamp method"**: Andy Warhol and Pat Hackett, *POPism: The Warhol Sixties* (Orlando: Harcourt, 1980), 22.

109 **"Warhol and his peers"**: Gérald Gassiot-Talabot, "Lettre de Paris," *Art International* 3, no. 2 (1964).

109 **"If silk-screening always evoked"**: Gopnik, *Warhol*, 251–52.

109 **"For me"**: Jerry Saltz, "Jerry Saltz on Andy Warhol's Portraits of Liz," *New York*, March 23, 2011.

110 **"It's impossible . . . reproduction"**: David Bourdon, interview with Andy Warhol, December 24, 1962, to January 14, 1963, in Kenneth Goldsmith, ed., *I'll Be Your Mirror: The Selected Andy Warhol Interviews, 1962–1987* (New York: Carroll & Graf, 2004), 6–14.

111 **"I was so jealous"**: Andy Warhol, *The Andy Warhol Diaries*, edited by Pat Hackett (New York: Warner Books, 1989), 586.

111 **"Would you like to replace human effort?"**: Gerard Malanga, "Andy Warhol on Automation: An Interview with Gerard Malanga," *Chelsea* 18 (1964), reprinted in Goldsmith, ed., *I'll Be Your Mirror*, 60–62.

112 *Life, Time, People,* and *Popular Mechanics*: "Andy's No-Man Show," *Life*, December 1984, photography by Eric Wexler; *Time*, November 15, 1982, photography by Eric Wexler; "Art Imitates Artist as Andy Warhol is Cloned," *People* 19, no. 18 (May 9, 1983); "Will the real Andy Warhol please stand up and say something?," *Popular Mechanics*, April 1984.

112 **"Andy loved this idea"**: David Buckley, *Kraftwerk: Publikation* (London: Omnibus, 2015), 179.

112 **"There was a reason"**: Author's interview with John Cale.

113 **more than sixty films**: See Bruce Jenkins and Tom Kalin, *The Films of Andy Warhol Catalogue Raisonné: 1963–1965*, edited by John Hanhardt (New York: Whitney Museum of American Art, 2021).

113 **"The technical capabilities"**: Author's interview with Steven Watson.

113 **"What you should do"**: Warhol, *Diaries*, 470.

114 **"Iggy's concept of the mechanical"**: Author's interview with John Cale.

CHAPTER 9: PATTERNS

117 **"There are no rules"**: Ted Loos, "Helen Frankenthaler, Back to the Future," *New York Times*, April 27, 2003.

117 **Helen Frankenthaler**: For more on Frankenthaler, see Alison Rowley, *Helen Frankenthaler: Painting History, Writing Painting* (London: I. B. Tauris, 2007).

117 **Grace Hartigan**: For more on Hartigan, see Cathy Curtis, *Restless Ambition: Grace Hartigan, Painter* (Oxford: Oxford University Press, 2015); Grace Hartigan, *The Journals of Grace Hartigan, 1951–1955*, edited by William T. La Moy and Joseph P. McCaffrey (Syracuse, NY: Syracuse University Press, 2009). For a consideration of both Frankenthaler and Hartigan, see Daniel Belasco, "See Us Now: The Feminist Positions of Helen Frankenthaler and Grace Hartigan, 1957–1962," *Kunsthistorisk Tidskrift / Journal of Art History* 83, no. 2 (2014).

118 **"I really loathe abstract expressionist painting"**: Hartigan, *Journals*, 30.

118 **"Computers accept instructions"**: Lejaren Hiller [Jr.], "Music Composed with Computers—A Historical Survey," in Harry B. Lincoln, ed., *The Computer and Music* (Ithaca, NY: Cornell University Press, 1970), 42.

119 **"The method is justified"**: Lejaren A. Hiller Jr. and Leonard M. Isaacson, *Experimental Music Composition with an Electronic Computer* (New York: McGraw-Hill, 1959), 68–69.

119 **"far closer to mathematics"**: Igor Stravinsky and Robert Craft, *Conversations with Igor Stravinsky* (Garden City, NY: Doubleday, 1959), 17. This comment is often misquoted as "Musical form is close to mathematics" or "Music is close to mathematics," shifting the meaning to make the statement more absolute.

119 **"all musicians are subconsciously mathematicians"**: Pearl Gonzalez, "Monk Talk," *Down Beat*, October 28, 1971.

119 twelve-tone and serial music: For more on the debates over twelve-tone and serial music, see Ray C. B. Brown, "Intelligibility First Consideration in Art: Composers Who Write in Idiom Known to Themselves Alone Incur Danger of Being Misunderstood by All but Themselves," *Washington Post*, February 21, 1937; Cyrus Durgin, "Interview with Symphony Guest Conductor: More About Ansermet's Views on Music in 12-Tone System," *Daily Boston Globe*, January 1, 1956; William Mayer, "Live Composers, Dead Audiences," *New York Times*, February 2, 1975; and Donal Henahan, "A Vote for Interpretation," *New York Times*, August 3, 1980.

120 a stack of four computer punch cards: To see photos of these cards, visit spalter digital.com.

120 **"his only thought"**: Gerald Jonas, "Op (Cont.)," in "The Talk of the Town," *The New Yorker*, February 27, 1965.

120 **"Wise was on the lookout"**: Author's interview with A. Michael Noll.

120 an article in *Scientific American*: Béla Julesz, "Texture and Visual Perception," *Scientific American* 212, no. 2 (February 1, 1965).

121 **"I told him about my work"**: Author's interview with A. Michael Noll.

121 vile, racist pseudo-science: For insight into Shockley's promotion of white supremacy and eugenics, see William Shockley, *Shockley on Eugenics and Race: The Application of Science to the Solution of Human Problems*, edited by Roger Pearson (Washington, DC: Scott-Townsend, 1992).

122 **"I don't know what Elwyn . . . got to work"**: Author's interview with A. Michael Noll.

123 **"The digital computer"**: A. Michael Noll, Memorandum to Bell Telephone Laboratories, "Patterns by 7090—Case 38794–23," August 28, 1962.

123 **"There was a legal issue . . . mechanism of randomness"**: Author's interview with A. Michael Noll.

124 **"This exhibition demonstrates"**: "Computer-Generated Pictures at the Howard Wise Gallery," press release, Howard Wise Gallery, April 6–24, 1965.

125 **"It possessed an insouciant elegance"**: Frank Gillette, "Howard Wise," *Provincetown Arts*, 1992.

125 **"The wave of the future"**: Stuart Preston, "Reputations Made and in Making," *New York Times*, April 18, 1965.

126 **"He foresaw the future"**: Achim Moeller, "Howard Wise Gallery: Exploring the New," Moeller Fine Art, April/July 2012.

126 **"Rules for action"**: Author's interview with Frieder Nake.

127 Max Bense: See Max Bense, *Aesthetica: Einführung in die Neue Aesthetica* (Baden-Baden: Agis, 1965). For a discussion in English of Bense's thoughts on information

aesthetics, see Christoph Klütsch, "Information Aesthetics and the Stuttgart School," in Hannah B. Higgins and Douglas Kahn, eds., *Mainframe Experimentalism: Early Computing and the Foundations of the Digital Arts* (Berkeley: University of California Press, 2012).

127 **"Bense, in a sense"**: Author's interview with Frieder Nake.

128 **"Principles of Generative Aesthetics"**: For more, see Max Bense, "The Project of Generative Aesthetics," in Jasia Reichardt, ed., *Cybernetics, Art and Ideas* (Greenwich, CT: New York Graphic Society, 1971).

128 **"No one understood a word"**: Author's interview with Frieder Nake.

128 **"It's nice and truly interesting"**: The description of the following encounter is drawn from the author's interview with Frieder Nake.

129 **"Klee became my source of inspiration"**: Author's interview with Frieder Nake.

130 **"Art does not reproduce the visible"**: Paul Klee, "Creative Confession, 1920," in Matthew Gale, ed., *Creative Confession and Other Writings* (London: Tate Publishing, 2013).

130 **"I was working in algorithms"**: Author's interview with Frieder Nake.

CHAPTER 10: SOME MORE BEGINNINGS

133 **"The Machine as Seen at the End of the Mechanical Age"**: Exhibition catalog and press materials can be seen at moma.org.

133 **"the mechanical machine"**: K. G. Pontus Hultén, *The Machine as Seen at the End of the Mechanical Age* (New York: Museum of Modern Art, 1968), 6.

134 **"We are surrounded"**: Ibid., 3.

134 **Hultén made international headlines**: See Richard Boston, "Hon," *New Statesman* 2 (July 1, 1966); Terry Coleman, "Strangest 'Woman' In the World," *San Francisco Examiner*, July 31, 1966.

135 **"historical collection of comments"**: Grace Glueck, "An Erotic Auto, A Roomful of Fog," *New York Times*, November 24, 1968.

135 **a bit about Aristotle**: Ibid., 6–7. Hultén devotes several paragraphs to a discussion of art, technology, and nature in Greek thought, quoting passages from the *Mechanica*, written by either Aristotle or his pupil Strato.

135 **"To contemporary spectators"**: Ibid., 21.

136 **"They posed a riddle"**: Barbara Gold, "A Look at What Happened to Art After the Machine," *Baltimore Sun*, December 1, 1968.

136 **"The industrialization"**: Hultén, *The Machine as Seen at the End of the Mechanical Age*, 173.

137 **"the oldest self-propelled vehicle"**: Ibid., 26.

138 **"Forces of construction"**: Ibid., 66.

138 **"the estrangement"**: Ibid., 127.

138 **"A thing which is perhaps"**: Christopher Andreae, "Art and the Machine," *Christian Science Monitor*, December 12, 1968.

139 **"Machines are beautiful"**: John Canaday, "Art: Machines Fascinate: Artifacts Upstage Their Human Aides—General Effect Is That of a Carnival," *New York Times*, November 28, 1968.

139 **"It takes a bit of doing"**: Ibid.

140 **"Some More Beginnings"**: For more, visit the website of the Brooklyn Museum. See also the exhibition catalog: Billy Klüver, Julie Martin, and Robert Rauschenberg, *Some More Beginnings: An Exhibition of Submitted Works Involving Technical Materials and Processes Organized by Staff and Members of Experiments in Art and Technology*

in Collaboration with the Brooklyn Museum and the Museum of Modern Art, New York (New York: Experiments in Art and Technology, 1968).

139 **"Maintain a constructive climate"**: Billy Klüver and Robert Rauschenberg, "Mission Statement," *Experiments in Art and Technology News* 2, no. 1 (March 18, 1968).

140 **"Like many works"**: K. G. Pontus Hultén, "Experiments in Art and Technology," press release, Museum of Modern Art, November 1968.

141 **Lillian Schwartz**: For more on the pioneering multimedia artist Lillian F. Schwartz and her work, see Lillian F. Schwartz and Laurens R. Schwartz, *The Computer Artist's Handbook: Concepts, Techniques, and Applications* (New York: W. W. Norton, 1992); Virginia Lee Warren, "An Artist Makes House Calls," *New York Times*, October 25, 1971; Alan M. Kriegsman, "Lillian Schwartz's Art by Computer," *Washington Post*, November 1, 1984; Lillian F. Schwartz, "The Computer and Creativity," *Transactions of the American Philosophical Society* 75, no. 6 (1985); and Gary Singh, "A Saturation of Firsts: Lillian F. Schwartz," *IEEE Computer Society*, September/October 2012. When contacted for this book, Laurens Schwartz declined to permit an interview with his mother, offering instead to speak for her on the condition that he be granted right of approval over any text pertaining to Lillian Schwartz, terms to which the author could not submit.

141 **"Mechanically, *Proxima Centauri*"**: Author's interview with Kenneth Knowlton.

141 **"Lillian was a fireball"**: Ibid.

142 **Three hundred sixty years**: For the conversion from the *Star Trek* dating system to our dating system, see hillschmidt.de/gbr/sternenzeit.htm.

143 **"Frankly, during the entire shooting"**: Leonard Nimoy, *I Am Spock* (New York: Hyperion, 1995), 115.

CHAPTER 11: IT'S LIKE A ROBOT

146 **"sanitizer"**: Albert Glinsky, *Switched On: Bob Moog and the Synthesizer Revolution* (New York: Oxford University Press, 2022), 148.

146 **Bernie Krause**: For more on Krause, his association with Robert Moog and Paul Beaver, and his other work, see Bernie Krause, *Into a Wild Sanctuary: A Life in Music and Natural Sound* (Berkeley, CA: Heyday, 1998).

147 **Harrison asked Krause to stay**: Ibid.

148 **"It was enormous"**: The Beatles, *The Beatles Anthology* (San Francisco: Chronicle, 2000), 340.

148 **"I want to play something . . . couple of quid"**: Krause, *Into a Wild Sanctuary*, 66.

149 **"I had no control"**: Graeme Thomson, *George Harrison: Behind the Locked Door* (London: Omnibus, 2013), 158.

149 **peaking at number 191**: See "Billboard 200," *Billboard*, July 12, 1969.

150 **"utter bullshit"**: Edmund O. Ward, review of John Lennon and Yoko Ono, *Unfinished Music No. 2: Life With the Lions* (Zapple ST-3357) and George Harrison, *Electronic Sounds* (Zapple ST-3358)," *Rolling Stone*, August 9, 1969.

150 **"quite well learning"**: Ibid.

150 **"THIS ALBUM"**: Loyd Grossman, "George Harrison: Electronic Sound (Zapple)," *Fusion*, August 8, 1969.

152 **"Leopold Stokowski was so inspired"**: Details on the Wanamaker organ demonstration come from the author's visit to Macy's, Philadelphia.

153 **"amazing and startling"**: "Capital Hears New Novachord Musical Device," *New York Herald Tribune*, February 2, 1939.

153 **"Most inventors"**: John Cage, *Silence: Lectures and Writings* (Middletown, CT: Wesleyan University Press, 1961), 3. According to Cage's introduction to the text in *Silence*, this was initially presented as a "talk at a meeting of a Seattle arts society organized by Bonnie Bird in 1937," and first printed in liner notes to the recording of Cage's music in 1958.

153 **"Its inventor"**: "If You Want to Make Music from Hot Trumpet to Violin, Organ Will Do It," *Austin American*, February 12, 1930.

154 **"We'll Meet Again"**: Vera Lynn, " 'We'll Meet Again,' Arthur Young (Novachord)," Michael Ross Ltd, 1939.

155 **made with 163 vacuum tubes**: For more on the Novachord and its history, see nova chord.co.uk.

155 **developing an electric organ**: For more on the history of Hammond organs, see Scott Faragher, *The Hammond Organ: An Introduction to the Instrument and the Players Who Made It Famous* (Milwaukee: Hal Leonard Corporation, 2011).

156 **"real," "fine," and "beautiful"**: "Federal Trade Commission Decision: Official Findings and Order," *American Organist* 21, no. 8 (August 1938).

156 **An impassioned new style**: For more on the history of soul music and the relationship between the sacred and the secular in African American music, see Joel Rudinow, *Soul Music: Tracking the Spiritual Roots of Pop from Plato to Motown* (Ann Arbor: University of Michigan Press, 2010); Johannes Riedel, *Soul Music, Black and White: The Influence of Black Music on the Churches* (Minneapolis: Augsburg, 1975).

157 **"We used the Moog"**: The Beatles, *The Beatles Anthology*, 340.

158 **Wendy Carlos**: For more on Carlos and her work, see Amanda Sewell, *Wendy Carlos: A Biography* (New York: Oxford University Press, 2020).

159 **"to demonstrate to the world"**: Albert Glinsky, *Switched On: Bob Moog and the Synthesizer Revolution* (New York: Oxford University Press, 2022), 137.

159 **"The whole record"**: Glenn Gould, presentation on *CBC AM*, CBC, November 1968.

159 **"We're just a baby"**: Wendy Carlos interviewed by Glenn Gould, *CBC AM*, CBC, November 1968.

160 ***Switched-On Bach* was an unexpected hit**: See "Billboard Top LPs," *Billboard*, April 26, 1969; "Col's 'Switched-on Bach' Tops Mil Copies Sold," *Billboard*, June 8, 1974.

160 **"It would be easy"**: Thomas Willis, "Records: Trans-Moog-rified Bach—Authentic Beat of Now," *Chicago Tribune*, December 8, 1968.

161 **"It's not one instrument"**: "Rock and Roll; Make it Funky; Interview with Malcolm Cecil and Robert Margouleff [Part 1 of 2]," GBH Archives, c. September 1995.

161 **"TONTO represented"**: Mark Mothersbaugh, liner notes to TONTO's Expanding Head Band, *TONTO Rides Again*, Viceroy Vintage, 1996.

161 **"He said, 'I don't believe' "**: Nelson George, *Where Did Our Love Go? The Rise and Fall of the Motown Sound* (New York: St. Martin's Press, 1985), 180.

162 **to record seventeen full tracks**: See Malcolm Cecil, "When Malcolm Cecil Met Stevie Wonder," redbullmusicacademy.com, April 7, 2014. Figures on the number of tracks Wonder, Margouleff, and Cecil recorded, as well as how long they recorded together, diverge in various accounts of their collaboration.

162 **"I was trapped"**: Stevie Wonder interviewed by Penny Valentine, *Sounds* 22 (January 1972).

162 **"Stevie was apparently"**: "Stevie's Wonder Men," presented by Stuart Maconie, BBC Radio 4, November 30, 2010.

163 **"I didn't know what he expected"**: "Rock and Roll; Make it Funky."

163 **"a way to directly express"**: Wonder interviewed by Valentine.

164 **"I love getting into"**: Ben Fong-Torres, "I Want to Get into as Much Weird Shit as Possible," *Rolling Stone*, April 26, 1973.
164 **"This collaboration"**: John Diliberto, "Interview with Malcolm Cecil," *Keyboard*, 1984, reprinted in Ernie Rideout, ed., *Synth Gods* (London: Backbeat, 2011).
164 **"Just look at what Stevie did"**: Author's interview with Quincy Jones.

CHAPTER 12: PARADISE

167 **German art-rock band Kraftwerk**: For more on Kraftwerk and their music, see Pascal Bussy, *Kraftwerk: Man, Machine, and Music* (London: SAF, 1993); David Buckley, *Kraftwerk: Publikation* (London: Omnibus, 2015).
168 **"We are playing"**: Chris Power, "Kraftwerk—The Man-Machine: Remastered," *Drowned in Sound*, October 13, 2009.
169 **"a cross between"**: Amanda Arber, "Classic Albums: Iggy Pop—The Idiot," clash music.com, March 16, 2012.
169 **"That was a very interesting moment . . . followed the technology"**: Author's interview with Giorgio Moroder.
171 **mistakenly describe "I Feel Love"**: See David Hajdu, *Love for Sale: Pop Music in America* (New York: Farrar, Straus & Giroux, 2016). On p. 184, I wrote, "As [Robert] Fripp explained, the producer Giorgio Moroder created the entire musical track electronically, without a single acoustic instrument."
172 **"one of the first [recordings]"**: Richard Vine, "A History of Dance Music: Donna Summer's 'I Feel Love,' " *Guardian*, June 15, 2011.
172 **"This is it—look no further!"**: Simon Reynolds, "Song from the Future: The Story of Donna Summer and Giorgio Moroder's 'I Feel Love,' " *Pitchfork*, June 29, 2017.
172 **"If any one song"**: Ibid.
172 *Generation Ecstasy:* Simon Reynolds, *Generation Ecstasy: Into the World of Techno and Rave Culture* (Boston: Little, Brown, 1998).
172 **"The sequencer bass"**: Don Snowden, "Moog on the State of the Synthesizer," *Los Angeles Times*, June 7, 1981. The quote followed a comment about "Star Cycle" by Jeff Beck. Moog said, "I notice the same sort of thing in the Donna Summer tune."
174 **The skill Nicky Siano tutored**: For more on the arts of DJing and turntabling, see Bill Brewster and Frank Broughton, *Last Night a DJ Saved My Life: The History of the Disc Jockey* (London: Headline, 1999).
177 **"At the Garage, I felt like"**: Christopher Vaughn, "Electric Man," unfinished memoir, 1987.
177 **"People would say"**: Jens Gerrit Papenburg, *Listening Devices: Music Media in the Pre-Digital Era* (New York: Bloomsbury, 2023), 147.
177 **"Kraftwerk were main components"**: Emmett G. Price III, Tammy Lynn Kernodle, and Horace Joseph Maxile, eds., *Encyclopedia of African American Music*, vol. 1 (Santa Barbara: ABC-CLIO, 2011), 405.
178 **"The kids that were hanging out"**: Frankie Knuckles interviewed by Frank Broughton, February 1995, DJ history archives, redbullmusicacademy.com.
178 **"If a song was 'house' "**: Brewster and Broughton, *Last Night a DJ Saved My Life*, 294.
179 **house was a right-brain art**: I use the term "right-brain" in the vernacular sense. Contemporary neuroscience recognizes the structure of the brain to be intricately complex and not bifurcated into diametric halves.
179 **"Rap and house"**: Don Snowden, "House Music: The Blues for Dance," *Los Angeles Times*, May 28, 1989.

179 **"My white coworkers"**: Vaughn, "Electric Man."

179 **"House records are like"**: John Leland, "House Music: Disco for a Complicated Time," *Newsday*, November 8, 1989.

180 **"The parties were very intense"**: Knuckles interviewed by Broughton.

180 **"So he came down"**: Ibid.

181 **"It was just classy"**: Mireille Silcott, *Rave America: New School Dancescapes* (Toronto: ECW Press, 1999), 26–27.

181 **"Within the last five years"**: S. Cosgrove, "Seventh City Techno," *The Face*, May 1988.

182 **"One day the robots"**: Pascal Bussy, *Kraftwerk: Man, Machine, and Music* (London: SAF, 1993), 107.

CHAPTER 13: A VERY CURIOUS RELATIONSHIP

186 **"Aaron is autonomous"**: David Holmstrom, "Programmed 'Arm' Rivals Artist's Hand," *Christian Science Monitor*, April 6, 1995.

186 **"It does things"**: Peter Monaghan, "Art Professor Creates a Computer That Draws and Paints," *Chronicle of Higher Education*, May 9, 1997.

186 **"Five Young British Artists"**: These artists were painters Bernard Cohen, Harold Cohen, Robyn Denny, and Richard Smith, and sculptor Anthony Caro.

186 **"defies description"**: Nigel Gosling, "Images of a Master," *Observer*, October 13, 1963.

186 **"Cybernetic Serendipity"**: See the exhibition catalog courtesy of Studio International online. See also María Fernández, "Detached from HiStory: Jasia Reichardt and *Cybernetic Serendipity*," *Art Journal* 67 (2008).

187 **Jasia Reichardt**: For more on Reichardt's thinking on art and technology, see Jasia Reichardt, *Cybernetics, Art, and Ideas* (Greenwich, CT: New York Graphic Society, 1971); Jasia Reichardt, "Machines and Art," *Leonardo* 20, no. 4 (1987).

187 **"Cybernetic Serendipity deals with possibilities"**: Jasia Reichardt, *Cybernetic Serendipity: The Computer and the Arts* (New York: Frederick A. Praeger, 1969), 5.

188 **"It cannot think"**: Barbara Gold, "The Cybernetic Serendipity World," *Sun* (Baltimore, MD), July 20, 1969.

188 **"either computers were awfully stupid"**: Pamela McCorduck, *Aaron's Code: Meta-Art, Artificial Intelligence, and the Work of Harold Cohen* (New York: W. H. Freeman, 1991), 31.

189 **"What I have done"**: Holmstrom, "Programmed 'Arm' Rivals Artist's Hand."

189 **"As human beings do"**: George Johnson, "In the Eye of the Beholder: What If a Piece of Art Has No Conscious Meaning?," *Los Angeles Times*, August 17, 1986.

190 **"What I intended"**: Harold Cohen, "The Further Exploits of Aaron, Painter," *Stanford Electronic Humanities Review* 4 (July 1995).

190 **"All its decisions"**: Ibid.

190 **"man–machine relationship"**: Harold Cohen, "On Purpose: An Enquiry into the Possible Roles of the Computer in Art," *Studio International* 187, no. 962 (January 1974).

192 **"a much better colorist"**: Martin Gayford, "Robot Art Raises Questions about Human Creativity," *MIT Technology Review*, February 15, 2016.

192 **"Art writers run"**: John Schwartz, "Is Aaron's Work Creative Art or Just High-Tech Doodling?," *Washington Post*, April 10, 1995.

192 **"The drawings"**: Grace Glueck, "Portrait of the Artist as a Young Computer," *New York Times*, February 20, 1983.

192 **"Aaron raises the question"**: Schwartz, "Is Aaron's Work Creative Art or Just High-Tech Doodling?"

193 "'**Creative' is a word**": Harold Cohen, "Colouring Without Seeing: a Problem in Machine Creativity," *AISB Quarterly* 102 (1999): 26–35.

193 "**Clearly, the machine**": Ray Kurzweil, *The Age of Intelligent Machines* (Cambridge, MA: MIT Press, 1990), produced with the Milton J. Rubenstein Museum of Science and Technology 1987 exhibition "Robots and Beyond: The Age of Intelligent Machines."

193 "**Aaron exists**": Cohen, "Further Exploits."

194 **If Cohen seemed elusive**: For a sampling of Harold Cohen's scholarly writing on Aaron and computer art, see Cohen, "Further Exploits"; Harold Cohen, "How to Draw Three People in a Botanical Garden," *AAAI '88: Proceedings of the Seventh AAAI National Conference on Artificial Intelligence*, August 21, 1988; and Harold Cohen, "Style as Emergence (from What?)," in S. Argamon, K. Burns, and S. Dubnov, eds., *The Structure of Style* (Berlin: Springer, 2010).

194 **Margaret A. Boden**: For a sampling of Boden's rich body of scholarship on creativity and artificial intelligence, see Margaret A. Boden, *Creativity and Art: Three Roads to Surprise* (New York: Oxford University Press, 2010); Margaret A. Boden, *Artificial Intelligence* (San Diego: Academic Press, 1996); and Margaret A. Boden, ed., *The Philosophy of Artificial Life* (New York: Oxford University Press, 1996).

194 "**Novelty is one . . . I like it**": Author's interview with Margaret A. Boden.

195 **Association for the Advancement of Creative Musicians**: For more on the AACM, see George Lewis, *A Power Stronger Than Itself: The AACM and American Experimental Music* (Chicago: University of Chicago Press, 2008). For more on George E. Lewis's thinking on technology and improvisation, see George E. Lewis, "Interacting with Latter-Day Musical Automata," *Contemporary Music Review* 18, no. 3 (1999). See also George E. Lewis, "Too Many Notes: Computers, Complexity, and Culture in Voyager," Anna Everett and John T. Caldwell, eds., *New Media: Theories of Practices of Digitextuality* (Abingdon, UK: Routledge, 2003).

196 **KIM-1 computer**: Stephen Edwards, professor of computer science at Columbia University, provided helpful insight into the KIM-1 computer.

199 "**This is where . . . We'll have to see**": Author's interview with George E. Lewis.

CHAPTER 14: TEACHING AND LEARNING

202 **can be registered for copyright**: In a touchstone ruling on intellectual property, *Burrow-Giles Lithographic Co. v. Sarony*, 1884, the Supreme Court decided that photographs qualify as works of authorship under US copyright law. The case involved an elaborately staged studio portrait of Oscar Wilde. Weighing whether or not the photo was "a mere mechanical reproduction," the court found the image to be "an original work of art, the product of plaintiff's intellectual invention, of which plaintiff is the author."

203 "**Computing Machinery and Intelligence**": Alan M. Turing, "Computing Machinery and Intelligence," *Mind* 59, no. 236 (October 1950).

203 "**Not until a machine**": Geoffrey Jefferson, "The Mind of Mechanical Man," *British Medical Journal*, June 25, 1949.

204 "**every aspect of learning**": J. McCarthy, M. L. Minsky, N. Rochester, and C. E. Shannon, "A Proposal for the Dartmouth Summer Research Project on Artificial Intelligence," August 31, 1955, 2.

204 "**The program knew**": Author's interview with George Lewis.

205 "**in a weird desperation**": Author's interview with David Cope.

205 **Over nearly fifty years:** For a sampling of David Cope's writing on music, computers, and computer music, see David Cope, *Computers and Musical Style* (Madison, WI: A–R Editions, 1991); David Cope, *Hidden Structure: Music Analysis Using Computers* (Middleton, WI: A–R Editions, 2008); and David Cope, *Virtual Music: Computer Synthesis of Musical Style* (Cambridge, MA: MIT Press, 2001).

206 **"It's all valid":** Author's interview with David Cope.

206 **For a class George Lewis once taught:** This account comes from the author's interview with George Lewis.

206 **"I think listeners":** Ibid.

208 **the AI categorized Black faces:** Andrew Hundt, William Agnew, Vicky Zeng, Severin Kacianka, and Matthew Gombolay, "Robots Enact Malignant Stereotypes," *FAccT '22: Proceedings of the 2022 ACM Conference on Fairness, Accountability, and Transparency* (New York: Association for Computing Machinery, 2022).

208 **"grew far faster":** Author's interview with Simon Colton.

208 **Simon Colton:** For more on Colton's thinking, see his scholarly writing on computational creativity and related topics: Simon Colton, "Creativity Versus the Perception of Creativity in Computational Systems," *AAAI Spring Symposium: Creative Intelligent Systems* (Palo Alto, CA: AAAI Press, 2008); Simon Colton and Geraint A. Wiggins, "Computational Creativity: The Final Frontier?," *ECAI 2012: 20th European Conference on Artificial Intelligence* (Amsterdam: IOS Press, 2012).

208 **the Painting Fool:** For more on the Painting Fool, see Simon Colton, "The Painting Fool: Stories from Building an Automated Painter," in J. McCormack and M. d'Inverno, eds., *Computers and Creativity* (Berlin: Springer, 2012).

209 **"learn about creativity":** Author's interview with Simon Colton.

209 **"My purpose":** Ibid.

209 **Tony Veale:** For a sampling of Veale's scholarship on artificial intelligence and computational creativity, see Philipp Wicke and Tony Veale, "Creative Action at a Distance: A Conceptual Framework for Embodied Performance with Robotic Actors," *Frontiers in Robotics and AI* 8 (April 30, 2021); Tony Veale, *Exploding the Creativity Myth: The Computational Foundations of Linguistic Creativity* (London: Bloomsbury Academic, 2012).

209 **"Can a machine move me":** Author's interview with Tony Veale.

210 **"Deep learning makes almost anything":** Author's interview with Simon Colton.

210 **"One of the reasons":** Ibid.

211 **"Gaining inspiration from human creative expression":** S. Colton, A. Pease, C. Guckelsberger, J. McCormack, and M. T. Llano, "On the Machine Condition and its Creative Expression," in *Proceedings of the International Conference on Computational Creativity*, Association for Computational Creativity, 2020.

211 **"This is going":** Author's interview with Simon Colton.

CHAPTER 15: ADVERSARIAL NETWORKS

217 **"Christie's continually stays attuned":** Naomi Rea, "Is the Art Market Ready to Embrace Work Made by Artificial Intelligence? Christie's Will Test the Waters This Fall," *Artnet News*, August 20, 2018.

217 **"when it goes under the hammer":** In response to this claim, *The Verge* reported a previous auction of AI art. See James Vincent, "How Three French Students Used Borrowed Code to Put the First AI Portrait in Christie's," *The Verge*, October 23, 2018.

218 **"The picture is a typical":** Jerry Saltz, "An Artwork Made by Artificial Intelligence

Just Sold for $400,000. I Am Shocked, Confused, Appalled," *New York*, October 25, 2018.

218 **"No algorithm can capture"**: Jonathan Jones, "A Portrait Created by AI Just Sold for $432,000. But Is It Really Art?," *Guardian*, October 26, 2018.

218 **"My reaction was"**: Gabe Cohn, "Up for Bid, AI Art Signed 'Algorithm,'" *New York Times*, October 22, 2018.

219 **"No one in the AI and art sphere"**: Tim Schneider and Naomi Rea, "Has Artificial Intelligence Given Us the Next Great Art Movement? Experts Say Slow Down, the 'Field Is in Its Infancy,'" *Artnet News*, September 25, 2018.

219 **"One thing about our art"**: Ciara Nugent, "The Painter Behind These Artworks Is an AI Program. Do They Still Count as Art?," *Time*, August 20, 2018.

219 **"Back then people were saying"**: Ibid.

219 **"by invading the territories of art"**: Charles Baudelaire, "On Photography," in Baudelaire's four-part essay "The Salon of 1859," *Revue Française*, June/July 1859.

220 **"Our purpose with AICAN"**: Author's conversation with Ahmed Elgammal at the HG Contemporary gallery.

220 **"an almost autonomous artist"**: Marian Mazzone and Ahmed Elgammal, "Art, Creativity, and the Potential of Artificial Intelligence," *Arts*, February 21, 2019.

221 **Marian Mazzone**: For more on Mazzone's scholarship with and without Elgammal, see Marian Mazzone and Ahmed Elgammal, "Artists, Artificial Intelligence and Machine-Based Creativity in Playform," *Artnodes*, no. 26 (2020); Marian Mazzone, "Andy Warhol: Computational Thinking, Computational Process," *Leonardo*, April 1, 2020.

221 **"What Ahmed did . . . in good ways"**: Author's interview with Marian Mazzone.

222 **"Our goal is to advance"**: Greg Brockman, Ilya Sutskever, and OpenAI, "Introducing OpenAI," Open AI blog, December 11, 2015.

222 **"We'll need to invest billions"**: Ibid.

222 **"While Jukebox represents"**: Prafulla Dhariwal, Heewoo Jun, Christine McLeavey Payne, Jong Wook Kim, Alec Radford, and Ilya Sutskever, "Jukebox," OpenAI website, April 30, 2020.

223 **"I think A.I. is great"**: Ezra Marcus, "The Future According to Grimes," *New York Times*, October 28, 2020.

225 **"I can see how A.I. art"**: Kevin Roose, "An A.I.-Generated Picture Won an Art Prize. Artists Aren't Happy," *New York Times*, September 2, 2022.

225 **"We're watching the death," "This isn't going to stop"**: Ibid.

226 **"There Is No Such Thing as A.I. Art"**: Walter Kirn, "There Is No Such Thing as A.I. Art," *Free Press*, September 25, 2022.

226 **"the exact opposite"**: Sarah Shaffi, "'It's the Opposite of Art': Why Illustrators Are Furious About AI," *Guardian*, January 23, 2023.

227 **"We have to try to understand"**: Author's interview with Simon Colton.

227 **"modern miracle"**: This account, and all quotes, come from the author's notes taken at the event.

Illustration Credits

Index

Page numbers in *italics* refer to illustrations. Page numbers after 232 refer to notes.